Living with intersex/DSD

Living with intersex/DSD

An exploratory study of the social situation
of persons with intersex/DSD

Jantine van Lisdonk

Netherlands Institute for Social Research | SCP
The Hague, August 2014

The Netherlands Institute for Social Research | scp was established by Royal Decree of March 30, 1973 with the following terms of reference:
a to carry out research designed to produce a coherent picture of the state of social and cultural welfare in the Netherlands and likely developments in this area;
b to contribute to the appropriate selection of policy objectives and to provide an assessment of the advantages and disadvantages of the various means of achieving those ends;
c to seek information on the way in which interdepartmental policy on social and cultural welfare is implemented with a view to assessing its implementation.

The work of the Netherlands Institute for Social Research focuses especially on problems coming under the responsibility of more than one Ministry. As Coordinating Minister for social and cultural welfare, the Minister for Health, Welfare and Sport is responsible for the policies pursued by the Netherlands Institute for Social Research. With regard to the main lines of such policies the Minister consults the Ministers of General Affairs; Security and Justice; the Interior and Kingdom Relations; Education, Culture and Science; Finance; Infrastructure and the Environment; Economic Affairs, Agriculture and Innovation; and Social Affairs and Employment.

© Netherlands Institute for Social Research | scp, The Hague 2014

scp publication 2014-23
Layout and design: Textcetera, The Hague
Translated from the Dutch: Julian Ross, Carlisle, uk
Cover design: Bureau Stijlzorg, Utrecht
Cover illustration: © Ien van Laanen, 2014

isbn 978 90 377 0717 5
nur 740

Distribution outside the Netherlands and Belgium : Transaction Publishers, New Brunswick (usa)

The Netherlands Institute for Social Research | scp
Rijnstraat 50
2515 xp Den Haag
The Netherlands
Tel. +31 70 340 70 00
Website: www.scp.nl
E-mail: info@scp.nl

The authors of scp publications can be contacted by e-mail via the scp website.

Contents

Foreword

It is a tradition in many countries, including the Netherlands, to take a gift for a newborn baby. In the Netherlands, the gift often indicates the sex of the baby: pink for a girl, blue for a boy. Uncommonly, it is not immediately clear what the sex of a baby is, or it turns out later that their sex development has proceeded differently from what is usual. There are many variations of this. The illustration on the frontispiece shows a few examples: boys with an extra X chromosome; girls with XY chromosomes; or babies whose sex at birth is difficult to establish. Intersex/DSD, as this phenomenon is known, can be related to chromosome variations, anatomy or the gonads.

In the light of the growing international political attention for human rights issues in relation to intersex/DSD, the Dutch Minister of Education, Culture and Science announced in an emancipation policy memorandum for the period 2013-2016 (*Hoofdlijnenbrief Emancipatiebeleid 2013-2016*) that intersex was being considered as a possible focus area for policy (TK 2012/2013). To date, the knowledge gained from research is limited mainly to insights obtained from a human rights or medical perspective. Little is currently known about the problems experienced by people with different types of intersex/DSD and precisely what issues they encounter in their lives. The Emancipation department at the Ministry of Education, Culture and Science asked the Netherlands Institute for Social Research|SCP to document what is currently known about the social situation of people with intersex/DSD. Such a publicly funded study is unique. The study was limited in scope and exploratory in nature.

All those who contributed to the study on the basis of their personal or professional experience and expertise by taking part in interviews and discussions did so with a remarkable degree of engagement. I would like to thank everyone sincerely for their openness and willingness to share their most personal experiences and life lessons. This demonstrates the need for recognition and illustrates the importance of this issue. The author wishes to thank the professional experts for their helpful comments on the sections of this report that are medical in nature, and also Joost Kappelhof for his important contribution to Appendix A. Great thanks are also due to the members of the reading committees and other colleagues for their commitment and valuable comments.

Prof. Kim Putters
Director, Netherlands Institute for Social Research | SCP

all of them have been through medical procedures and are or have been in contact with patient organisations.

Variation, prevalence, medical interventions and terminology

There is considerable variation in the conditions that are covered by the term 'intersex/ DSD'. What they have in common is that they are congenital and incurable. There are dozens of conditions, some of which occur in many variants. Conditions may be discovered at different moments. Some of them may be discovered during pregnancy, for example in the case of chromosome variations. If the intersex/DSD affects the external genitalia, the condition is generally identified at birth. Where there are consequences for the internal sex organs, this is frequently only discovered during puberty or adulthood. Conditions sometimes come to light when a woman is unable to become pregnant. It is also possible for people not to know that they have an intersex/DSD condition, for example because their medical complaints are limited or a link is not made to intersex/DSD. The nature and severity of the medical consequences also vary greatly. The medical complaints may be so minor that medical treatment is never required, whereas others with the condition face a whole lifetime of medical interventions and monitoring. Those interventions may include operations on the internal or external sex organs, hormone therapy or treatment of additional physical complaints. Operations are sometimes needed because otherwise there is a danger to a child's life; operations are also frequently carried out in the case of ambiguous genitalia or where the genitals are incompletely formed or developed. Many types of intersex/DSD have an impact on fertility, but advancing medical techniques are increasing the possibilities for fulfilling the desire to have children, for example through surrogacy, egg cell donation or artificial insemination.

It is difficult to indicate how often intersex/DSD occurs in the Netherlands. This depends among other things on the classification chosen and the prevalences applied per condition. In some cases, reliable prevalences are simply not available. If we take the most generally accepted classification among medical professionals, non-medical researchers and interest groups as a starting point, and if we also as far as possible start from the available condition-specific prevalences in the Netherlands, the best estimate at this point in time is that there are roughly 80,000 people in the Netherlands with a type of intersex/DSD. That number includes people who have never received medical treatment for their condition and people who are not aware that they have a type of intersex/DSD.

Personal experiences

It emerged from the interviews with persons with intersex/DSD that many of them had (and sometimes still have) difficulty in accepting their condition. For all those interviewed, receiving the news of their condition was a very distressing event; coming to terms with having a chronic condition and its physical consequences can be difficult. Types of intersex/DSD which affect the person's external appearance can also impact on their self-image as a man or woman. The persons interviewed know perfectly well whether they feel male or female, but sometimes wonder whether others see them as a complete man or woman. Infertility can also lead to great sadness, both for the

Foreword

It is a tradition in many countries, including the Netherlands, to take a gift for a newborn baby. In the Netherlands, the gift often indicates the sex of the baby: pink for a girl, blue for a boy. Uncommonly, it is not immediately clear what the sex of a baby is, or it turns out later that their sex development has proceeded differently from what is usual. There are many variations of this. The illustration on the frontispiece shows a few examples: boys with an extra X chromosome; girls with XY chromosomes; or babies whose sex at birth is difficult to establish. Intersex/DSD, as this phenomenon is known, can be related to chromosome variations, anatomy or the gonads.

In the light of the growing international political attention for human rights issues in relation to intersex/DSD, the Dutch Minister of Education, Culture and Science announced in an emancipation policy memorandum for the period 2013-2016 (*Hoofdlijnenbrief Emancipatiebeleid 2013-2016*) that intersex was being considered as a possible focus area for policy (TK 2012/2013). To date, the knowledge gained from research is limited mainly to insights obtained from a human rights or medical perspective. Little is currently known about the problems experienced by people with different types of intersex/DSD and precisely what issues they encounter in their lives. The Emancipation department at the Ministry of Education, Culture and Science asked the Netherlands Institute for Social Research|SCP to document what is currently known about the social situation of people with intersex/DSD. Such a publicly funded study is unique. The study was limited in scope and exploratory in nature.

All those who contributed to the study on the basis of their personal or professional experience and expertise by taking part in interviews and discussions did so with a remarkable degree of engagement. I would like to thank everyone sincerely for their openness and willingness to share their most personal experiences and life lessons. This demonstrates the need for recognition and illustrates the importance of this issue. The author wishes to thank the professional experts for their helpful comments on the sections of this report that are medical in nature, and also Joost Kappelhof for his important contribution to Appendix A. Great thanks are also due to the members of the reading committees and other colleagues for their commitment and valuable comments.

Prof. Kim Putters
Director, Netherlands Institute for Social Research | SCP

Summary

Intersex/DSD is an umbrella term used to describe various congenital conditions
in which the development of sex differs from what medical professionals generally
understand to be 'male' or 'female'. The differences may be chromosomal, gonadal or
anatomical.
It is not so much the person's perception of their own sex that is the issue here; people
with a type of intersex/DSD almost always feel that they are male or female.

In the interviews we conducted for this study, it quickly emerged that the terminology
is a highly sensitive area. There is a wide difference between medical professionals on
the one hand and human rights organisations and advocates on the other. In this report
we use the term 'intersex/DSD'. The term 'intersex' is mainly used by human rights organisations and advocates, and is also the usual term used in the political domain. DSD,
standing for *disorders of sex development*, is generally the preferred term among medical
professionals. An alternative definition of DSD is *differences of sex development*; that term is
preferred by people who consider the word 'disorders' to be pejorative. In general, intersexuality and hermaphroditism are now regarded as outdated and pejorative terms.

Intersex/DSD from the perspective of social situation

Attention for intersex DSD was for a long time largely confined to medical circles. However, attention for intersex/DSD is today growing internationally in the context of the
protection of human rights, embracing issues of self-determination and physical integrity in relation to genital operations and irreversible medical interventions performed
on children. At the same time, there is little solid, research-based knowledge, and it is
unclear what impact intersex/DSD has on the social situation of those affected.
At the request of the Emancipation department of the Dutch Ministry of Education,
Culture and Science, the Netherlands Institute for Social Research | SCP carried out an
exploratory study in order to gain a greater insight into the social situation of people
with intersex/DSD. There were two central research questions:
1 What characterises the social situation of people with intersex/DSD and what problems do they experience in their social situation?
2 How could a reliable quantitative study of the social situation of people with intersex/
   DSD be carried out in the Netherlands?

'Social situation' is a broad term; in this report, we distinguish between the personal
experience of intersex/DSD and the impact of intersex/DSD on a person's social life.
This is an exploratory study with limited scope, based on national and international
literature; seven in-depth interviews with persons with intersex/DSD; a focus group
discussion with people having various types of intersex/DSD who are active in patient or
advocacy groups; eight interviews with professional experts; and several meetings and
symposia. The persons interviewed with intersex/DSD constitute a specific group, since

all of them have been through medical procedures and are or have been in contact with patient organisations.

Variation, prevalence, medical interventions and terminology

There is considerable variation in the conditions that are covered by the term 'intersex/ DSD'. What they have in common is that they are congenital and incurable. There are dozens of conditions, some of which occur in many variants. Conditions may be discovered at different moments. Some of them may be discovered during pregnancy, for example in the case of chromosome variations. If the intersex/DSD affects the external genitalia, the condition is generally identified at birth. Where there are consequences for the internal sex organs, this is frequently only discovered during puberty or adulthood. Conditions sometimes come to light when a woman is unable to become pregnant. It is also possible for people not to know that they have an intersex/DSD condition, for example because their medical complaints are limited or a link is not made to intersex/DSD. The nature and severity of the medical consequences also vary greatly. The medical complaints may be so minor that medical treatment is never required, whereas others with the condition face a whole lifetime of medical interventions and monitoring. Those interventions may include operations on the internal or external sex organs, hormone therapy or treatment of additional physical complaints. Operations are sometimes needed because otherwise there is a danger to a child's life; operations are also frequently carried out in the case of ambiguous genitalia or where the genitals are incompletely formed or developed. Many types of intersex/DSD have an impact on fertility, but advancing medical techniques are increasing the possibilities for fulfilling the desire to have children, for example through surrogacy, egg cell donation or artificial insemination.

It is difficult to indicate how often intersex/DSD occurs in the Netherlands. This depends among other things on the classification chosen and the prevalences applied per condition. In some cases, reliable prevalences are simply not available. If we take the most generally accepted classification among medical professionals, non-medical researchers and interest groups as a starting point, and if we also as far as possible start from the available condition-specific prevalences in the Netherlands, the best estimate at this point in time is that there are roughly 80,000 people in the Netherlands with a type of intersex/DSD. That number includes people who have never received medical treatment for their condition and people who are not aware that they have a type of intersex/DSD.

Personal experiences

It emerged from the interviews with persons with intersex/DSD that many of them had (and sometimes still have) difficulty in accepting their condition. For all those interviewed, receiving the news of their condition was a very distressing event; coming to terms with having a chronic condition and its physical consequences can be difficult. Types of intersex/DSD which affect the person's external appearance can also impact on their self-image as a man or woman. The persons interviewed know perfectly well whether they feel male or female, but sometimes wonder whether others see them as a complete man or woman. Infertility can also lead to great sadness, both for the individual concerned and for those close to them. Parents with a child who has a type of intersex/DSD also have to go through a process of self-acceptance and coming to terms with the reality. Finally, intersex/DSD can have an impact on health and well-being because of condition-specific aspects (including physical complaints, problems with physical appearance), medical interventions and social treatment (which can exacerbate feelings of being different, somehow inferior and unhealthy) and actual or anticipated reactions from others in society.

Intersex/DSD and the social environment

Most persons with intersex/DSD who were interviewed find it difficult to be open about their condition. In practice, they decide to be cautious about whom they tell and what they tell them. They do not always reveal everything and they sometimes adapt or avoid certain situations in order to ensure that their condition remains hidden. Their condition often has (or has had) an impact on their ability to form relationships and on their experience of sexuality. They may for example feel uncertainty about their appearance or their self-image as a man or woman, reticence in embarking on a relationship, fear of rejection as a partner and limitations in their sexual capabilities. Discovering that they may be infertile can also be difficult to come to terms with and impact on their ability to have children.

People with intersex/DSD may have feelings of shame and a fear of negative reactions in relation to those around them. Those reactions appear to stem mainly from ignorance, embarrassment and an inability to empathise. The interviewees with intersex/DSD seem able to understand such reactions to an extent and generally do not associate them directly with non-acceptance or discrimination, though they can cause them to feel different, lonely and misunderstood. They are much less accepting of negative reactions and ignorance from the medical profession, because they expect medical professionals to have adequate knowledge and expertise and to treat them with respect.

This study shows that physical limitations and reduced psychosocial well-being can have an adverse impact on participation in education, work and leisure activities. Gaining an impression of how extensive that impact is would require additional research. People appear to be especially vulnerable during puberty, because this is a time when sexuality, intimacy and appearance are sensitive subjects. However, our study is based on the experiences of adults who recount their experiences in puberty retrospectively, which means that any conclusions about this phase of life need to be treated with some caution.

The experiences of children, adolescents and parents are not covered in this study.

All interviewees with intersex/DSD reported that contact with others with a similar condition is important in enabling them to share experiences and information based on a sense of equality, or not feeling isolated or different. The need for this contact is sometimes temporary. In the Netherlands, people with intersex/DSD organise themselves into separate patient organisations for each condition. There is currently no umbrella organisation in the Netherlands focusing on intersex persons in general or persons with intersex/DSD who do not regard themselves as patients.

Pointers for policy development

Intersex/DSD: lack of group identity and relationship with LGBT

Interest groups representing lesbian, gay, bisexual and transgender persons (LGBT), as well as human rights organisations, are increasingly adding the I of 'intersex' to the abbreviation LGBT. This implies that there are clear communalities between people with an intersex condition and people who are gay, lesbian, bisexual or transgender. What they have in common is that they encounter normativity and preconceptions in relation to sex, gender and sexuality. An alliance between LGBT and 'I' is therefore understandable. However, this study shows that such an alliance is a sensitive subject for persons with intersex/DSD. In the first place, there is virtually no shared identity or community based on intersex/DSD. These persons generally do not feel part of a group and do not wish to be seen as a separate category, but rather as men and women. In addition, most of the interviewees in this study prefer to distance themselves from LGBT persons or feel they have little in common with them. They fear that they themselves will also be viewed in terms of sexual orientation or gender identity, whereas these two aspects are not crucial for them. Persons with intersex/DSD for whom gender identity and gender expression are more ambiguous may feel some connection with LGBT persons, but the information on this is limited.

It is important that any development of policy on intersex/DSD takes account of these sensitivities and the lack of a group identity on the part of those with these conditions. We would also note that in a policy approach focusing on impediments that have their origin in normativity, sensitivities and perceptions in relation to sex, gender and sexuality, there are clear correspondences with women's emancipation and LGBT emancipation.

Better picture of social situation through research

This exploratory study identifies a number of themes that are relevant for people's social situation, but there are still gaps in the knowledge. If there is an interest in developing policy in this field, it is important to increase that knowledge. Among other things, large-scale quantitative research could be carried out aimed at obtaining a national picture of the social situation of persons with intersex/DSD. The opportunities and difficulties have been explored in this study. This study devoted little attention to questions of medical ethics, whereas this can be relevant for policy development.

Fostering resilience through organisations

The persons with intersex/DSD who were interviewed for the study recounted that they had not found it easy to obtain good information and support. Self-organisation in the form of condition-specific patient organisations appears to play an important role in fostering the resilience of persons with intersex/DSD, as a means of finding information, meeting others and learning to come to terms with their own condition and to deal with those around them. On a practical level, some patient organisations are not easily able to guarantee their continuity and the availability of activities. A self-organised body for persons who identify themselves as intersex or an activist movement has not been identified in the Netherlands (there are examples of both in other countries). Whether or not there is a need for such organisations in the Netherlands was not covered in this study.

Good information is an important condition for resilience. Although the Internet is a source of information, it is difficult to separate the good information from the bad, and the information is sometimes also pejorative and sensationalist. It is therefore no surprise that there is an expressed need for a website containing good information about the medical and social aspects of types of intersex/DSD.

Finally, most of those interviewed have or have had a need for psychosocial support. For various reasons, they have not always received this.

Improving knowledge and sensitivity among medical professionals

There have been many changes in the medical world in recent decades, and developments in this field are rapid. One result of this is that previous practices such as secrecy and deliberately withholding information or providing false information are no longer standard practice. Within specialist medical centres, full disclosure is now the norm. Despite these developments, it became clear in this study that there is generally room for improvement in the knowledge, awareness and sensitivity of non-specialist medical professionals. Since most persons with intersex/DSD initially come into contact with non-specialists, this suggests a need to devote attention to intersex/DSD in basic medical training and training for the care professions. The drive to improve knowledge moreover needs to focus on both the medical and social aspects of intersex/DSD, sex and sexuality. Another reason for raising the sensitivity of professionals is that being treated with respect is likely to foster a positive experience of medical intervention. Intersex/DSD is surrounded by sensitivities in relation to sex and sometimes external appearance, infertility and sexuality, and medical professionals need to be more aware of this. Provision of adequate information by medical professionals to persons with intersex/DSD is also something that warrants attention; the interviewees with intersex/DSD felt they had received inadequate information from doctors and were left with many questions. They often use the Internet and patient organisations as supplementary sources of information.

Finally, non-medical researchers and advocates argue that medical professionals need to devote more attention to views on a person's sex and to the framing of intersex/DSD as a medical problem. Medical intervention in the event of ambiguous genitalia, for example, confirms the norm of a person's sex as being a clear-cut twofold division. What is regarded as 'normal sex' within the medical profession is always based on biological and social criteria, neither of which are fixed.

Perceptions, acknowledgement, invisibility and awareness-raising

An important finding of this exploratory study is that there is a need for medical and social acknowledgement. There is a great deal of ignorance about intersex/DSD. The persons with intersex/DSD who were interviewed for this study come up against this ignorance and, with some conditions, against the perception of a person's sex as a simple dichotomy in which the distinction male/female is taken for granted as the norm (sex normativity). Intersex/DSD differs from that norm. There are taboos and sensitivities here because it impinges on issues in relation to sex and sexuality.

Persons with intersex/DSD generally do not feel the need to question the male/ female distinction, and in most cases prefer to be seen as a complete man or woman. This means that intersex/DSD is not very visible in society. Whether more visibility would be beneficial or should be a goal is a complex matter. Advocates believe that striving for this will have a positive effect because it could lead to less ignorance, more nuanced perceptions, greater familiarity with the condition and the breaking of taboos. Others fear that greater visibility could lead to more stigmatisation. In addition, greater visibility and awareness in society could lead to intersex/DSD being seen as a distinct social category or group, whereas most of those interviewed do not want this.

Promoting sharing of knowledge

Finally, this study showed that there are wide differences between the various parties and organisations involved, such as medical professionals, advocates and patient organisations, in terms of their perspectives, views and use of terminology in relation to intersex/DSD. Those differences encompass language usage, type of knowledge, relevant themes and the way in which intersex/DSD is viewed. The Netherlands Network Intersex/ DSD (Nederlands Netwerk Intersekse/DSD, NNID), a new intersex/DSD advocacy group that was set up last year, has organised several meetings in which various organisations have participated. This sharing of perspectives and information between parties and organisations, whatever form it takes, has proved to be useful.

1   Introduction

1.1   Intersex/DSD unknown among the general public

'Is it a boy or a girl?' That is often one of the first questions asked when a new baby is born. The answer is usually clear, but it occasionally happens that doctors are unable to determine the baby's sex immediately, or that they have doubts. The development of secondary sex characteristics in puberty can also be incomplete or not take place, and there may be a different chromosome pattern from xx or xy. These are examples which illustrate that a baby's sex at birth is not always unambiguously 'male' or 'female'. These people have an intersex condition[1] or DSD (*differences/disorders of sex development*). In this report we refer to these conditions using the term 'intersex/DSD' (see § 2.1).[2] There are many types of intersex/DSD; it is an umbrella term. It can include people whose external appearance is feminine but who have ambiguous or incompletely developed internal sex organs. It can also happen that the sex organs are not typically male or female at birth, or that females have no xx chromosomes and males no xy chromosomes. Some intersex/DSD conditions are visible to other people, while others are not. Some conditions are identified prior to or immediately after birth, while others only become apparent during puberty or following investigation when a woman is unable to become pregnant. People with intersex/DSD generally do not see themselves as a 'third gender', but as male or female. The most accurate assessment that can be made at present is that there are approximately 80,000 people in the Netherlands with a form of intersex/DSD (§ 2.5). These people are not always aware of their condition themselves. Intersex/DSD is a largely unknown phenomenon among the general public. People who have never personally come into contact with intersex/DSD often think of the term 'hermaphrodites'; this is an outdated and pejorative term used to describe people with both male and female genitalia. Then there is the attention devoted by the media to female athletes who were banned from participating in the Olympic Games because, according to medical standards, they were allegedly not completely female. At the end of 2013, a change in the law in Germany making it legally possible to have an indeterminate sex led to an appearance on the Dutch TV talk show *Pauw en Witteman* by someone who regards herself as intersexual. These examples show that intersex/DSD is not only a physical condition with medical consequences, but also a social issue which is associated among other things with pejorative perceptions and possible impediments to social participation.

The impact of intersex/DSD varies, and depends on when it is discovered by medical professionals, how and when people are informed about it, external appearance, the severity of any physical complaints and the social treatment of others. Intersex/DSD is sometimes established before birth, and a person with intersex/DSD may spend their whole lives under medical treatment (possibly with many operations). But there are also people who will never discover that they have intersex/DSD, because the medical symptoms are minor or are not recognised.

What these types of intersex/DSD have in common is that from a medical perspective they all point to some variation or ambiguity in sex development. Almost nothing is known about what impact this has on the social situation of people with intersex/DSD. There is growing international political attention for this (European Parliament 2014; European Union 2013; Council of Europe 2013; Méndez, United Nations 2013), but as yet little solid knowledge from research. This exploratory study marks a first step towards gaining a greater understanding of the social situation of people with intersex/DSD.

1.2   Intersex/DSD: an emancipation issue?

For decades, attention for intersex/DSD has been largely limited to a medical perspective. Although the emphasis still lies on medical aspects, there is growing attention for and recognition of the importance of the non-medical aspects (MacKenzie et al. 2009; Vilain 2006). Human rights organisations and advocates of people with intersex/DSD are raising questions about issues such as self-determination and physical integrity in relation to genital operations and irreversible medical interventions performed on children. Some organisations and scientists are also raising the more fundamental question of whether a person's sex should be interpreted as an absolute dichotomy (male or female), when intersex/DSD demonstrates that this is not always clear-cut (Karkazis 2008; Liao & Boyle 2004b; Van Heesch, forthcoming).

The international attention devoted to intersex/DSD from a human rights perspective has recently been stepped up. General human rights organisations such as the United Nations (Méndez, United Nations 2013) and European organisations such as the Council of Europe (2013), the European Union (2013) and the European Parliament (2014) have recently begun focusing on the position of people with intersex/DSD. New legislation has also come into being in several countries in recent years in relation to sex registration and antidiscrimination (e.g. Germany, Australia and South Africa).

The Dutch Minister with responsibility for emancipation policy is monitoring international developments and in an emancipation policy memorandum tabled in Parliament for the period 2013-2016 (*Hoofdlijnenbrief Emancipatiebeleid 2013-2016*) announced that the possibility is being explored of making intersex as a potential policy domain:

> *Recent publications and statements have focused on combating discrimination and protecting physical integrity and self-determination, because intersexual people may among other things face unnecessary medical interventions. In consultation with relevant patient organisations, I am exploring whether there are societal issues here and whether these have an emancipation aspect.* (TK 2012/2013)

1.3   Design and implementation of exploratory study

Before the question can be answered of whether including intersex/DSD within the emancipation policy is useful and necessary, it is important to have an understanding of the problems faced by people with intersex/DSD. What are their lives like and what issues do they encounter? This exploratory study seeks to answer these questions from the perspective of persons with intersex/DSD, with a focus on their social situation. It looks at which problems stem from having an intersex/DSD condition. Issues relating to human rights and medical ethics fall outside the scope of this study and are mentioned in the main text only if interviewees with intersex/DSD broached them themselves. A number of human rights issues are referred to in Appendix C.

'Social situation' is a broad term, which among other things embraces social acceptance, well-being and health, and participation. In this report we distinguish between personal experiences of intersex/DSD and social interaction with other people. With regard to personal experiences, we look at how people with intersex/DSD experience their condition (e.g. medical interventions, identification, self-acceptance, self-image as a man or woman, consequences for health and well-being). As regards social interaction with others, we look at their openness about their own condition, relationship and family formation, social treatment of others, perceived acceptance and discrimination, and difficulties or impediments in participating in education, work and leisure activities.

This study also explores the question of whether there is a need for large-scale quantitative research on the social situation of people with intersex/DSD and how such research could be carried out. We identify *which* problems occur in the social situation of people with intersex/DSD, but are unable to elucidate *to what extent* these problems occur; that would require national research involving a large group of people with intersex/DSD. Such research has never been carried out in the Netherlands. Research among people with intersex/DSD from a non-medical perspective is also scarce in other countries (MacKenzie et al. 2009). Almost nothing is therefore known about how large-scale quantitative research on the social situation of people with intersex/DSD might be carried out and what the focus areas, challenges and opportunities might be.

Summarising, the following two questions are central to this exploratory study:

1   What characterises the social situation of people with intersex/DSD and what problems do they experience in their social situation?
2   How could a reliable quantitative study of the social situation of people with intersex/DSD be carried out in the Netherlands?

To answer these two research questions, the study draws on four information sources: a national and international literature review; interviews with persons with intersex/DSD; interviews with professionals working in the field of intersex/DSD; and meetings and symposia. However, this study is limited in scope and is no more than an exploration of the field. This means that the literature review is not exhaustive and that the number of interviews and discussions, while it provides an insight into the identified themes, does not provide an in-depth picture of aspects such as differences between groups. Moreover, information was only obtained about people who are aware of their condition.

The literature review covers social scientific studies, medical science literature (to gain an insight into medical issues and prevalences), websites of Dutch patient organisations and websites and reports of interest groups (in particular the Netherlands Network Intersex/DSD (NNID), COC (Dutch lesbian, gay, bisexual, transgender organisation) and ILGA (International Lesbian, Gay, Bisexual, Trans and Intersex Association)).

In-depth face-to-face interviews were held with seven persons with various conditions that are classified as intersex/DSD. All persons interviewed have at some point received treatment in a medical centre because of their condition and are or have been in contact with a patient organisation or support group. Most of them are or have also been active in patient organisations and therefore have some insight into the experiences of others with their condition. In addition, a focus group discussion was held involving six people with various types of intersex/DSD who are active in patient organisations or interest groups. They talked about their own experiences and those of people in their organisations or in their social networks. The interviewees can be regarded as the 'vanguard'. This group is selective because not everyone with intersex/DSD undergoes medical treatment or has (had) contact with patient organisations. This needs to be taken into account when interpreting the findings.

Eight interviews were held with professional experts, including a social science researcher, representatives of the NNID, medical professionals and psychologists. A number of informal discussions also took place.

A number of international and national symposia were attended for the purposes of the study. In addition, four meetings were organised by the NNID in 2013 and 2014 in collaboration with the Ministry of Education, Culture and Science, which were attended by representatives of patient organisations, expertise centres, the COC, researchers and other stakeholders.[3] The organisation of these meetings was prompted by the founding of the NNID in the spring of 2013. This organisation campaigns for the equal treatment, emancipation, visibility and interests of people with intersex/DSD. It is the first organisation to focus specifically on promoting the interests of people with intersex/DSD in the Netherlands.

Given the social sensitivity of intersex/DSD, those who participated in the interviews were guaranteed anonymity. Names were changed; the only interviewees mentioned by name are professional experts or organisations that wished to be named.

1.4   Structure of this report

Chapters 2, 3 and 4 of this report address the first research question about social situation. The second research question is highly technical in nature and is therefore discussed (at length) in Appendix A.

Since intersex/DSD is an unknown phenomenon for many people, chapter 2 first presents a medical explanation of the condition, following which we describe some of the best-known types of intersex/DSD. We also explain the terminology and definitions we have used and look in more depth at how frequently intersex/DSD occurs. Chapter 3 discusses the relevant themes in relation to personal experience of intersex/DSD, while chapter 4 explores the consequences for social interaction with others. In order to bring the persons concerned to life, boxes in chapters 3 and 4 home in on specific life experiences or problems faced by the interviewees.

Appendix A discusses the second research question, namely how reliable quantitative research on the social situation of people with intersex/DSD could be carried out. After exploring the utility of and need for such research, we suggest a number of themes and address methodological questions. We then outline a number of specific possibilities and describe the advantages and disadvantages of each.

In the concluding discussion we summarise the findings on the two research questions. The role of perceptions, the dilemma of visibility and focus areas for the medical profession are also covered here. We also show where the gaps in knowledge are and whether and how quantitative research could contribute to filling them. Finally, we interpret the findings of this exploratory study in the light of potential policy development.

Comparisons are occasionally made in this report with former or current emancipation processes in relation to inequality between men and women and the acceptance of LGBT persons (lesbian, gay, bisexual and transgender persons). This enables us to draw parallels with the emancipation of other groups, or to highlight differences.

One outcome of this exploratory study is that there are many perspectives from which answers can be sought to the research questions There is wide variation in people's experiences with and views on intersex/DSD. Certain themes proved to be highly relevant for specific types of intersex/DSD but to play no role whatsoever for other conditions. There was also frequently wide divergence in the perspectives of persons with intersex/DSD, patient organisations, interest groups, researchers and medical professionals. We have attempted to strike a balance in portraying those perspectives.

Notes

1   In the Netherlands the term 'condition' is used by persons with intersex/DSD, advocates and medical professionals. It is more neutral than 'disorder' or 'abnormality' (see chapter 2). In other countries, the term 'condition' is generally not used by activists and human rights organisations, because they associate it with a medical approach. The situation appears to be different in the Netherlands, where the term 'condition' (*conditie*) receives more support.

2   Where reference is made to other authors or organisations, we have as far as possible used their terminology.

3   Those present were informed about national and international developments in the field of human rights and medical ethics. Knowledge of and experiences with various topics were also shared.

2 What is intersex/DSD? Types, terminology and prevalence

In order to gain a better understanding of the problems that may occur in the social situation of people with intersex/DSD, it is important to be clear about what intersex/DSD is. This chapter begins with a medical explanation and a description of the best-known types of intersex/DSD. This is followed by a discussion of terminology, definitions and demarcation of the target group. There are sensitivities here, and we highlight these briefly because they show how people who are personally or professionally involved with intersex/DSD view this phenomenon. We then show that intersex/DSD is linked to a person's sex and describe the relationship with sexuality, gender identity and sexual orientation. Finally, we attempt to answer the question of how frequently intersex/DSD occurs.

2.1 Types of intersex/DSD: a medical explanation

Although the medical approach is not central to this study, Box 2.1 presents a concise medical explanation of 'normal' sex development in order to illustrate how many physical processes underlie that development. This makes it possible to place different types of intersex/DSD more into context. Some processes are presented in simplified form.

Box 2.1 Brief medical explanation of normal sex development

Sex development begins at conception, when the genetic material from an ovum merges with the genetic material from a sperm cell and a new individual begins to develop. The sex chromosomes X and Y are important in sex development. The gonads develop during the first six weeks of the embryo's development. Initially, they are the same in XY and XX individuals; under normal circumstances, these gonads develop between week seven and week ten into testes or testicles in XY individuals, which produce androgens ('male' hormones). If the foetus has XX chromosomes, the gonads almost always develop into ovaries which produce oestrogen ('female' hormones). Once the testes have developed, they start producing testosterone and anti-mullerian hormone (AMH). AMH suppresses the development of female internal sex organs, i.e. the fallopian tubes, uterus and upper portion of the vagina. If the testes are functioning properly and AMH is present, the development of the male internal sex organs is initiated: epidymides, ductus deferens and seminal vesicles. Development of the external genitals takes place between week nine and week 14. Testosterone stimulates the growth of a penis and scrotum. If no testosterone is produced, or if it cannot function for some reason, the external genital organ generally develops into a vagina. The sex hormones testosterone and oestrogen initiate the development of secondary sex characteristics during puberty. The production of testosterone by the testes causes the voice to break and the development of body hair and more muscle development. Oestrogen, which is produced by the ovaries, stimulates the development of breasts, rounded hips and menstruation. Hormones also have an influence on behaviour and psychological functioning in puberty and in later stages of life.

The description in Box 2.1 makes clear that sex development is determined by a combination of chromosomes, gonads (testes or ovaries) and the production and processing of hormones. If one of these processes develops in a different way, this can lead to a type of intersex/DSD. These are congenital conditions for which there is no cure. There are dozens of types of intersex/DSD, often with many variations, which are highly diverse in terms of how and when they are discovered, the type of medical interventions performed and the medical and social consequences. In general, intersex/DSD which affects the external sex organs is discovered at birth. Types which have affected the internal sex organs are often only discovered during puberty or adulthood. Surgery may be performed in the case of ambiguous genitals. Operations are sometimes medically necessary, because otherwise the children concerned would not survive. There is a wide diversity of types of intersex/DSD. Here we describe a few of the best-known types.

46,XY-DSD

46,XY-DSD covers several types of intersex/DSD, including androgen insensitivity syndrome (AIS) and gonadal dysgenesis. People with AIS have XY chromosomes and are partially or completely insensitive to androgen. They have no ovaries or uterus. They may have little or no underarm hair, acne or body odour, and they may have pale nipples. AIS occurs in different gradations, and a distinction is made between CAIS (complete) and PAIS (partial). People with CAIS almost always feel themselves to be women and their external genitals look female. CAIS mostly comes to light during investigations as to why a girl is not menstruating during puberty. It is sometimes diagnosed earlier because of an inguinal hernia or because it is known to run in the family. People with PAIS have ambiguous external genitalia and this is discovered immediately after birth. Operations may be performed on the external sex organs of people with PAIS after extensive evaluation. These persons are always infertile. Because of the risk of cancer developing in the internal sex organs, these are sometimes removed. If the gonads are not properly developed, there may be another type of intersex/DSD, namely complete or partial gonadal dysgenesis. Partial forms of testosterone biosynthesis defects such as 5alpha RD-2 and 17beta HSD-3 also come under 46,XY-DSD.

Congenital adrenal hyperplasia (CAH)

In the most common form of congenital adrenal hyperplasia (CAH), the adrenal glands produce too little cortisol and too much testosterone. In girls, the clitoris begins growing even before birth and can resemble a small penis, while the labia may be completely or partially fused together. This can lead to problems in determining the child's sex. Girls with this condition often undergo operations on the external sex organs in early childhood. These girls display cross-gender (in this case masculine) behaviour more often than other girls, but generally still feel themselves to be women. If not treated properly, the high testosterone levels can later cause girls to develop extensive body hair and extreme acne. If left untreated, an increase in the 'masculine' hormones leads to early onset of puberty in both girls and boys. Boys with CAH are often ultimately smaller than their peers.

The milder forms of CAH are discovered in female children at the toddler stage due to the early onset of puberty, or in later life because of more pronounced hair growth, acne or menstrual disorders. They are discovered in males due to fertility problems caused by the slight increase in the 'masculine' hormones produced by the adrenal gland. Women with CAH can become pregnant and have children, but require careful medication. Narrowness of the vagina or early operations often means that a Caesarean section is needed for delivery.

Today, CAH is included in the neonatal screening test, which means the condition comes to light earlier than in the past.

Hypospadias

Hypospadias affects boys and means that the external opening of the urethra (urinary meatus) is not located at the tip of the penis. It may be located anywhere along the length of the penis or in the scrotum, and the foreskin is usually split. If the opening is closer to the scrotum, there is a greater chance of a curved penis. The position and shape of the urethral opening can make it difficult to urinate whilst standing. One or more operations can make it possible for these boys to pass water whilst standing and enable ejaculation later in life. The appearance of the penis can also be surgically altered. These operations are normally carried out at an early age. Medically, it is not always necessary to treat hypospadias, and the decision to operate is sometimes based mainly on practical and cosmetic considerations. Boys with hypospadias are fertile. The degree of hypospadias can vary widely, and the more severe forms are classed as DSD.[1]

MRKH (Mayer-Rokitansky-Küster-Hauser) syndrome[2]

MRKH occurs only in females. They have xx chromosomes and ovaries, but are born without a (complete) vagina and/or uterus. The external sex organs are present and these women feel themselves to be women. MRKH is usually discovered during puberty, because menstruation fails to begin or problems arise with the kidneys. If a (partial) uterus is present, it is usually removed to relieve the severe abdominal pain caused by the inability to menstruate. These women are not fertile. Surrogacy is an option.

Klinefelter syndrome

This syndrome, also known as 47,XXY, only affects babies born as boys. They have an extra X chromosome in addition to the usual 46 chromosomes. The syndrome affects roughly 1 in 600 or 700 men, making it one of the most common types of intersex/DSD. Males with this syndrome produce insufficient testosterone, leading to late onset of puberty and sometimes failure to complete puberty. The penis and testes are relatively small prior to puberty. The consequences of insufficient testosterone production become apparent during puberty. Affected boys develop relatively less muscle mass and the bodily distribution of fat and hair is often different from most of their peers. There may be some breast formation. The testes remain underdeveloped in adulthood. This syndrome may be accompanied by a list of symptoms, including problems with concentration, memory, coordination, fatigue, expression of emotions and social interaction. Those affected rarely display all symptoms and there are wide individual differences. As a result, Klinefelter syndrome is not always recognised and diagnosed by doctors. It may be that boys do not go to a doctor because they have few complaints, or it may be that complaints are not properly diagnosed. Once the diagnosis has been established, they are often given testosterone in the form of a gel, injection or capsule. This sometimes gives them more energy, improves their concentration, gives them more interest in sex, makes them more assertive, helps them develop muscle mass and alters the fat distribution around their body. It also reduces the risk of osteoporosis. These men often have fertility problems because of the rapid decline in sperm cell production during puberty.

Most boys who grow up with Klinefelter syndrome feel themselves to be men in adulthood, though in some cases they begin to feel more feminine or identify themselves as women.

Turner syndrome[3]

Turner syndrome is also known as 45,X. There are mosaic forms of the syndrome where the person affected has 45,X in some cells and, for example, 46,XX or 46,XY in others. Only females are affected and in all cases they lack all or part of one X chromosome. That affects the development of ovaries, the production of sex hormones, their height and physical sexual maturity. Affected girls and women are almost always short in stature. As this symptom is accompanied by a list of possible symptoms, which almost never all occur at the same time in one individual, there are wide differences in the medical issues faced by those affected. What they have in common is that the onset of puberty is generally not spontaneous and that they usually do not menstruate. This syndrome is often identified during puberty. Other examples of possible symptoms are a wide and/or short neck, low hairline at the back of the neck, an increased likelihood of heart problems, kidney, thyroid gland and hearing disorders, and more common incidence of diabetes, high blood pressure, reduced motor development and deficiencies in spatial awareness and short-term memory. Oestrogen may be administered to initiate menstruation and breast formation. Pregnancy is possible via egg cell donation.

Other chromosome variations

There are other chromosome variations in addition to Klinefelter syndrome and Turner syndrome, including 47,XYY and 47,XXX. Both these chromosome variations usually go undetected at birth. Boys with 47,XYY have an extra Y chromosome. In the early years of life, there is generally nothing unusual. At the end of primary school age, these boys put on an extra spurt of growth and often end up taller than average. Compared with other boys, they may be somewhat more mobile and active and have a greater chance of having learning difficulties and delayed emotional development. Males with this condition are fertile.

A chromosome variation that affects girls is 47,XXX, also referred to as triple-X syndrome or Trisomy X. Girls affected may have problems with language or speech, have socio-emotional problems, be taller than average and have a more negative self-image. They may be slightly more mobile and active than other girls. They are fertile and have a slightly higher probability of early menopause.

Both 47,XYY and 47,XXX are often not diagnosed and no medical intervention takes place, either because the condition does not lead to major problems or because problems are not associated with a chromosome variation. There are also mosaic patterns in which chromosome variations occur only in some cells.

Clearly, the experiences of people with different types of intersex/DSD vary widely as regards medical intervention (operations and hormone treatment) and fertility. People's personal experiences are discussed in chapter 3.
What intersex/DSD makes clear is that a person's sex is not an absolute dichotomy medically speaking, even though this is the most widely prevailing view of sex in society. It is more accurate to imagine a person's sex as a continuum on which there is a gradation of medically determined male or female characteristics.

2.2 Sensitivities surrounding terminology: intersex and/or DSD

It became apparent at an early stage of this exploratory study that the choice of terminology is a crucial issue within this field, and one which played a role in gaining access to individuals and organisations for this study. Broadly speaking, professionals prefer the umbrella term intersex or DSD or some variation of it. There is lively debate among scientists, advocates and medical professionals, because the term chosen reveals the perspective from which this group is viewed (Liao & Roen 2014).
DSD, in its most commonly used meaning of *disorders of sex development*, is currently the most usual term in national and international medical circles (e.g. Callens 2014; Claahsen-Van der Grinten et al. 2008; Dessens & Cohen-Kettenis 2008; Hughes et al. 2006; Van der Zwan 2013). Prior to 2006, medical professionals used the term intersex, but later came to see this as pejorative, partly because of its supposed implication that those affected were somewhere between male and female, whereas in reality most of them feel clearly male or female. The publication of the Consensus Statement in 2006 marked the reaching of a consensus in the international medical world on the use of the term *disorders of sex development*. This also led to a broad consensus regarding which conditions are regarded as DSD as well as the names given to specific conditions henceforth (Hughes et al. 2006).
However, human rights organisations, advocates and people with intersex/DSD, in particular, often object to the use of the term DSD, which they feel places too much emphasis on the medical approach (e.g. Karkazis 2008; NNID 2013). The criticism is directed mainly at the first 'D', in the sense of *disorders*. Not everyone with intersex/DSD feels like a patient, and terms such as 'disorder' or 'abnormality' are regarded as pejorative. The preferred terms of these organisations are intersex and persons with an intersex condition (e.g. Agius & Tobler 2012; COC 2013; ILGA Europe 2013; Karkazis 2008; MacKenzie et al. 2009; Council of Europe 2013).[4]
The lack of consensus on terminology sometimes stands in the way of cooperation between interest groups, patient organisations and medical professionals.[5] In order to facilitate dialogue with all organisations in the field, some organisations now use the term DSD but give it an alternative meaning, namely *differences of sex development*

(e.g. NNID 2013; Swiss National Advisory Commission 2012), *diverse sex development* (EUROP-SI 2014; Liao & Simmonds 2014), *divergence of sex development* (Reis 2007) or an alternative form such as *variation of sex development* (e.g. Van Heesch (forthcoming); NNID 2013; Swiss National Advisory Commission 2012).

As stated in chapter 1, in this exploratory study we use the terms 'intersex/DSD' and 'persons with intersex/DSD'.[6] These terms have been chosen in order to indicate that we do not adopt a fundamental position in the debate about the terminology. Our primary aim is to present different perspectives rather than (inadvertently) excluding some groups because of the choice of a particular terminology. We have nothing to gain if people refuse to cooperate because of our choice of words. This pragmatic approach proved to be necessary (see note 4).

Finally, it transpired that interviewees with intersex/DSD themselves use the usual terminology in very different ways. Some regard themselves as intersex or refer to DSD as a diagnostic term, but they are the exception; interviews with active members of patient organisations and persons with intersex/DSD revealed that they mostly do not use either term themselves. Instead, they generally tend to use the condition-specific term, such as 'men with Klinefelter syndrome' or 'women with XY chromosomes'. Some of those interviewed were actually found to be entirely unfamiliar with the terms intersex and DSD.

The terms 'hermaphrodite' and 'intersexual' are less commonly used today by medical professionals, advocates and affected persons themselves; see Box 2.2.

Box 2.2   Hermaphrodite and intersexuality: outdated and pejorative

The terms 'hermaphrodite' and 'intersexuality' are today regarded as outdated, confusing and pejorative (Reis 2007). The term 'hermaphrodite' refers to a person with both male and female sex characteristics. There is criticism of this terminology because the determination of a person's sex then depends to a large extent on the anatomy of the gonads, whereas other factors such as genes, chromosomes, hormones and gender identity also play a role (Vilain 2006). The terms 'intersexuality' or 'intersexual' are still used occasionally (e.g. Deutscher Ethikrat 2012; Vilain 2006), but are often seen as socially undesirable because they imply that the condition is about sexuality, which it is not. The concept 'intersexuality' also suggests that a person *is* intersexual, whereas many persons with intersex/DSD perceive their condition as something that they *have*.

2.3   Definitions

There is also no single definition of intersex/DSD that is endorsed by all interest groups, patient organisations and medical professionals. We will look briefly here at some of the definitions that are used in the literature in certain circles.

The definition of DSD as set out in the Consensus Statement (Hughes et al. 2006) and which is widely used in international medical science and among specialist medical professionals in the Netherlands is as follows:

The term 'disorders of sex development' (DSD) is proposed, as defined by congenital con-ditions in which development of chromosomal, gonadal, or anatomical sex is atypical.
(Hughes et al. 2006: 554)

The range of conditions covered by this definition and included in the same Consensus Statement (Hughes et al. 2006) is broadly accepted both within and outside the medical profession (e.g. Callens 2014; NNID 2013; Van der Zwan 2013).
Interest groups, human rights organisations and scientists generally follow the Consensus Statement by taking the same factors that are relevant for sex development. The difference is that they do not regard this sex development as 'atypical' in persons with intersex/DSD. The psychologists Liao and Boyle, for example, refer to intersex as

The outcome of sexual differentiation processes resulting in the person's chromosomal, go-nadal or genital characteristics not clearly corresponding to one of our categories of male and female (2004a: 446).

They therefore imply that 'male' and 'female' are not fixed biological givens, but are categories defined by people. The Dutch advocacy groups NNID and COC adopt a comparable definition of intersex/DSD:

Intersex is an umbrella term used to describe various congenital conditions in which the development of chromosomal, gonadal or anatomical sex differs from the norm. (NNID 2013: 21; COC 2013: 10)

The definition adopted in this report is based on all the above definitions:

Intersex/DSD is an umbrella term used to describe various congenital conditions in which the development of sex differs from what medical professionals generally understand to be 'male' or 'female'. The differences may be chromosomal, gonadal or anatomical.

In the first place, this definition combines intersex and DSD to show that both terms relate to sex development. The possible differences are the same as in the definitions referred to above. This definition shows that intersex/DSD and norms regarding sex development and what is generally considered 'male' or 'female' are not absolute. The definition is therefore not based on a person's perception of their own sex. As stated, persons with a form of intersex/DSD almost always feel male or female.[7]

2.4 Differences between intersex/DSD and transgender persons and sexual orientation

As intersex/DSD is sometimes confused with being transgender or with being gay or lesbian, in this section we explain the difference between these groups. Put briefly, intersex/DSD is related to a person's sex, not to their sexuality, gender identity or sexual orientation.
As a condition, intersex/DSD is not directly related to sexuality, though it can have an impact on how someone perceives and experiences sexuality. Women with a shallow vagina, men with a small penis and people who have undergone many operations on their external genitals may for example feel more uncertain in sexual contact and embarking on sexual relationships (Callens 2014; Van Den Dungen et al. 2002). A lot of

medical psychology research measures satisfaction with medical intervention by the degree to which persons with intersex/DSD have sexual contact and find it a pleasurable experience (e.g. Callens 2014; Cohen-Kettenis 2010). The impact of intersex/DSD on sexual and relational experiences is discussed in chapter 4.

The distinction between intersex/DSD and gender identity is also important. Intersex/DSD affects a person's *body* and sex development. Those affected often feel completely a man or a woman, and this corresponds with their sex at birth as determined by medical practitioners. They are male or female in appearance and their gender identity – the feeling of being a men or a women – is generally congruent with that appearance. This is not the case with transgender persons; they experience a discrepancy between their sex at birth and their *feeling* of being a men or a woman (Keuzenkamp 2012). This occurs sporadically with specific types of intersex/DSD (Cohen-Kettenis 2010), for example in persons with CAH and PAIS. It is also known that children with the most severe form of CAH or with PAIS are relatively often more masculine in their behaviour and interests than the average for girls (Cohen-Kettenis 2010; Dessens & Cohen-Kettenis 2008).[8] There are also examples of people with Klinefelter syndrome who grow up as boys but who later come to feel more feminine or that they are a woman.

Like the rest of the population, most people with intersex/DSD are heterosexual. There is generally no direct relationship between intersex/DSD and sexual orientation (i.e. to which sex a person feels sexually and romantically attracted). Here again, there are exceptions for specific conditions. For example, girls with certain forms of CAH feel attracted to members of their own sex slightly more often than average (Meyer-Bahlburg et al. 2008).

2.5 Prevalence of intersex/DSD

As intersex/DSD is a relatively unknown phenomenon, it is often assumed that it is not very common. However, it is not easy to give a good and above all unambiguous picture of the prevalence. In the first place, this is because of choices made in deciding what is classified as intersex/DSD. In addition, several prevalence figures may be in circulation for different conditions of intersex/DSD. Prevalences of specific conditions sometimes differ widely from one country to another (e.g. 5alpha-RD2 and 17beta-HSD3). There are also several conditions for which good, reliable prevalences are not available. As far as we are aware, Dutch medical practitioners and psychologists specialising in DSD have never published a total prevalence for DSD in the Netherlands. The question of how often DSD occurs is a pertinent one from a societal perspective, and we will therefore attempt to answer this question for the Dutch population. To this end we present four total prevalences, stating in each case which classification has been used and which source the prevalence derives from.

Table 2.1
Classification of four total prevalences, Dutch population (in percentages)

source of prevalence	scope	classification	population
known estimate of ambigu-ous genitals 1:4500, cited e.g. in Hughes et al. 2006	international	limited to ambiguous genitals	0.022
overview study Blackless et al. 2000	international	broad classification	1.7
NNID advocacy group, own calculation	Dutch, supplemented by international	based on Consensus Statement; calculation built up as prevalence of ambiguous genitals + nine other conditions	0.6
SCP, own calculation	Dutch, supplemented by international	based on Consensus Statement; calculation built up as prevalence of ambiguous genitals + eight other conditions	0.5

Source: SCP

Based on the most restrictive classification, in which the classification of intersex/DSD is narrowed to include only forms of ambiguous genitals, a prevalence of 1 in 4,500, or 0.022% of the population, both internationally and in the Netherlands, is generally accepted by medical professionals (e.g. Callens 2014; Hughes et al. 2006; Van der Zwan 2013). These are persons who cannot be immediately determined as male or female at birth (Vilain 2006). Although Dutch medical professionals cite this prevalence in studies, they also note that it is uncertain whether the international prevalence constitutes an accurate estimate for the Netherlands. Most professionals in the field of intersex/DSD regard a classification of the condition which includes only ambiguous genitals at birth as too restrictive, because other (possibly) severe types of intersex/DSD such as MRKH, CAIS, Klinefelter syndrome and Turner syndrome are excluded.
The other three prevalences cited here are based on or correspond with the generally accepted classification in the Consensus Statement (Hughes et al. 2006). That classification is widely accepted by medical practitioners, non-medical researchers and Dutch interest groups, and is also used as a basis in this study.
The highest estimate that receives broad support comes from the international overview study by Blackless et al., which reports a prevalence of 1.7% of the population (2000). The classification of intersex/DSD in this broad-based international prevalence includes all conditions in which there is a difference of sex development based on chromosomes, gonads, genitals and hormone levels and for which prevalences are known. The condition-specific prevalences are generally based on a number of studies. The Dutch

advocacy groups NNID (2013) and COC (2013) take this broad international prevalence figure as an upper limit.

Given the wide regional differences found for a number of conditions,[9] it is important when determining the total prevalence to make as much use as possible of studies carried out on the Dutch population. In its document *Standpunten en Beleid* ('Standpoints and Policy') (2013), the NNID attempted to arrive at an estimate for the Netherlands based on the classification in the Consensus Statement. It assumes a prevalence of 0.6% in the Dutch population, though the NNID adds the caveat that reliable condition-specific prevalences are not known for the Netherlands for all conditions, so that the actual total prevalence of the entire range of intersex/DSD conditions in the population could be higher. The calculation is the sum of condition-specific prevalences of ambiguous genitals;[10] 47,XXY (Klinefelter syndrome); 45,X (Turner syndrome); 47,XYY; 47,XXX; CAIS; Swyer syndrome; MRKH; hypospadias; and micropenis.

The Netherlands Institute for Social Research|SCP has calculated a prevalence indication for the Dutch population. It draws on the medical sources that underpin the prevalence estimate of the NNID, on Dutch medical publications and on information on patient organisation websites. Where no condition-specific prevalences were known for the Netherlands, international sources were used. SCP attempted to draw on at least two sources for each condition. The calculation includes the same conditions as in the total prevalence figure published by the NNID, with the exception of micropenis, since it is known that there is a good deal of overlap between this condition and other conditions classed as intersex/DSD. As with the NNID's estimated prevalence, only the severe form of hypospadias is included.[11] This leads to an estimated total prevalence of approximately 0.5% of the Dutch population. We should stress here that this prevalence indication is a global estimate based on existing medical sources. We would also add the caveat that specific prevalences are not known for every condition that is included in the classification used in the Consensus Statement. We were also unable to determine the degree of overlap between conditions. The total prevalence indication also includes people with intersex/DSD who are not aware of their condition,[12] for example people with 47,XXY, 47,XYY or 47,XXX. It is difficult to determine the size of this group. We also do not know how many people have been diagnosed or have at some point received medical treatment. Finally, we would note that roughly half the total prevalence indication percentage relates to boys and men with severe hypospadias (see Appendix B for more information on the calculation and the underlying sources).[13]

In conclusion, we can state that there is no uniform consensus regarding how often intersex/DSD occurs, and that any estimate depends greatly on the classification used and on the conditions for which prevalences are available. At this point in time, the best possible estimate of the prevalence of intersex/DSD in the Dutch population is around 0.5%; that equates to something over 80,000 persons, in the same range as the estimated prevalence of transgender persons in the Netherlands (Kuyper 2012).[14]

Notes

1 In calculating the prevalence of intersex/DSD, both SCP and NNID included only the severe forms of hypospadias; see also note 11.

2 The term MRK (without the H) is also often used in the Netherlands.

3 At the time of publication of this report, the patient organisation Turner Contact Nederland announced that they have doubts as to whether they should be classified under intersex/DSD. Members of the organisation feel that a label is being forced upon them which they do not recognise. They see points of contact, but above all wide differences compared with other conditions that are classified as intersex/DSD.

4 In Germany and a few other countries, the term inter* is used in a comparable way to trans*, where the asterisk can be replaced with any chosen term. This use of inter* in the Netherlands is highly unusual.

5 There are activists in some other countries who associate DSD so strongly with a medical approach, to which they are opposed, that they are unwilling to cooperate with organisations which use that term (with whichever meaning). There are also organisations in the Netherlands which prefer not to use the term DSD because they do not want people with these conditions to be seen as 'patients'. On the other hand, it is generally acknowledged that people with intersex/DSD sometimes need medical treatment. The term 'intersex' met with resistance from some Dutch medical professionals , who strongly advised using only the term DSD in the future. People who have not been diagnosed with DSD may call themselves intersex, and those with an intersex condition are consequently regarded as a diffuse group. Intersex is also sometimes seen as a non-medical, sociological concept, and for some people it is too politically loaded. Finally, the connotation with sex and sexuality is regarded as undesirable (Reis 2007). In some interviews with medical professionals it became apparent that they regard the use of the term intersex as a denial of the fact that those concerned have a medical condition and need treatment.

6 It would be more correct to speak of persons with an intersex condition/DSD, but for the sake of legibility it was decided to use a shorter variant.

7 People sometimes also see and identify themselves as being intersex, but have not been diagnosed as such by medical practitioners.

8 Interviewees with CAIS and Turner syndrome stated that women within their patient organisation very definitely feel themselves to be women and are often very feminine in their gender expression (verbal interviews).

9 This is the case, for example, with genetic conditions and conditions that are more likely to occur in cases of consanguinity (relationships between persons who share an antecedent). Some types of intersex/DSD occur more than average among migrant groups where endogamous marriages are common. An example is CAH.

10 These include PAIS, CAH, partial gonadal dysgenesis and partial forms of testosterone biosynthesis defects (e.g. 5alpha-RD2 and 17beta-HSD3) (NNID 2013).

11 There is a good deal of discussion about whether hypospadias should be regarded as a type of intersex/DSD, because the nature and severity of the condition can vary greatly (Hughes 2010). In the Consensus Statement, severe forms of hypospadias do count as DSD (Hughes et al. 2006). In our discussions with Dutch medical professionals, too, we found that there was a lack of consensus about whether or not hypospadias should be classified under DSD. The social science researcher

Van Heesch argues that hypospadias should be regarded as intersex because those affected often face operations and may be stigmatised because their penis looks different from the average (verbal interview, 19 December 2013). In this prevalence figure, hypospadias in de vicinity of the glans is left out of consideration, because this is regarded as a mild form of the condition (Pierik et al. 2002; NNID 2013).

12 Some condition-specific prevalences were determined by screening the number of liveborn babies in a given period, following which a calculation was carried out for the population as a whole. Consequently, it is possible that a prevalence of a condition is higher than the percentage in the population who are themselves aware of the condition.

13 Condition-specific prevalences were only included in the calculation if there was access to the original scientific articles explaining the calculation of the prevalence.

14 Kuyper observes that between 0.4% and 0.8% of Dutch men and between 0.1% and 0.3% of Dutch women report having an ambiguous or incongruent gender identity, experiencing feelings of gender dysphoria and having a desire for medical treatment (2012).

3   Personal experiences of intersex/DSD

In this chapter we look at people's personal experiences of intersex/DSD. What is it like for someone to have intersex/DSD? How do they experience their condition and how does it affect them? First, we describe the various ways in which people discover their condition. Since this is a condition which may entail medical aspects, we explore people's experiences of medical interventions and medication. We look at this in some detail because it shows how intersex/DSD manifests itself and how different the nature of medical interventions for conditions classed as intersex/DSD can be. We then look at the extent to which people have difficulty in coming to terms with their condition and whether the condition has consequences for their self-image as a man or woman. A central theme in most interviews was that people have difficulty accessing good information and obtaining support. That is also discussed here, as is how people ultimately do find the information they are looking for. We conclude with a résumé of what is known about the impact of intersex/DSD on perceived health and well-being.

The exploration of themes and issues is based mainly on aspects that emerged in the seven individual interviews and the focus group session with persons with intersex/DSD. Where possible, this is supplemented with information from a literature review, interviews with experts in the field and a number of informal conversations.

This chapter devotes a lot of space to presenting the views of interviewees with intersex/DSD. However, we would stress once again that these are persons who have at some time received medical treatment and who have (had) contact with patient organisations. We do not have a clear picture of persons who have no experience of this or who are not aware of their intersex/DSD condition.

3.1   The discovery and the message

People can discover that they have intersex/DSD in very different ways. With some conditions, it is clear immediately after birth that a baby has a form of intersex/DSD, because the appearance of the external genitals differs from the medical norm; the baby may for example have a very enlarged clitoris or fused labia. Another example is when the urethra of a baby boy is positioned in such a way that medical intervention is necessary immediately after birth. Chromosome variations are today identified much earlier thanks to prenatal diagnostics, enabling parents to be informed of the presence of conditions such as Turner syndrome and Klinefelter syndrome before the child is born. It is also possible that a girl fails to begin menstruating after entering puberty and that medical investigation reveals a form of intersex/DSD. Failure to become pregnant, finally, can also result in the discovery of intersex/DSD. In the past, doctors believed that it was better not to inform patients about their condition because the impact might be too much for them. Today, full disclosure is the norm.

One thing that all conditions have in common is that their discovery comes as a great shock. All interviewees with intersex/DSD were able to recount in great detail how they discovered their condition. We present a few examples from the interviews in this chapter, because they provide a good illustration of the impact, both for the individual concerned and for those close to them. They also make clear that the taboo and embarrassment can be considerable for doctors and/or parents.

Most of those interviewed were advised of their condition following medical investigation because of the failure of puberty to begin, failure to become pregnant or because of new medical complaints in adulthood. In one case, the interviewee's condition was discovered immediately after birth and a medically necessary operation was carried out straightaway.

The way in which the news is delivered is important for the way in which it is received. Some interviewees felt that the doctors who informed them of their condition lacked empathy, or else had empathy but delivered the news clumsily, making it more traumatic. A few examples are given below.

Els, a woman in her forties, underwent chromosome tests and screening for Turner syndrome at the age of 13 due to the lack of any sign of puberty beginning. Her mother was first informed by the doctor. The doctor's intention was to provide reassurance, but Els' mother experienced it differently:

> Els: *What he wanted to say to reassure her [was]: 'They almost all end up married, you know.' At which my mother, who was sitting down, stood up again: 'Is that supposed to reassure me?!' [Laughs]. My mother knew so little. There was no Internet or anything like that in those days. My mother had read something in a medical encyclopaedia, and that doesn't do anything to cheer you up. She didn't know if I had a normal life expectancy or serious medical problems. And then he comes out with that line... It was well intentioned, but it was ...*
>
> [...]
>
> *[The doctor] clearly thought it was important that I should be told. He sat there trying to persuade my mother. They knew then that I had Turner syndrome, but I didn't know it yet. But you end up in the medical system, so there was a heart examination, kidney testing, and all the rest of it.*
>
> [...]
>
> *We were walking down the corridor and he asked me: 'Do you know why you've just had these tests?' And I said, well, they found something with my bladder in an earlier test, and they want to investigate further ... That's what my mother had told me. But I saw from his reaction that my mother hadn't told me everything or that [it] didn't quite add up. So my mother told me afterwards, sitting in the car in the car park of the [name of medical centre].*
>
> Interviewer: *And what did she tell you?*
>
> Els: *Well, what she knew. That I didn't... I don't know exactly what she said. The main thing I remember was that I wouldn't be able to have children. I can't remember exactly what else she told me. But that, ...that was the thing.*
>
> Interviewer: *And how did that feel? Because that's a very different message from being told that there's something wrong with your bladder.*
>
> Els: *Yes, it felt awful. It was really distressing. Hearing something like that, because, well, as a girl you are of course the most...I was already very into in all of that, just like almost all girls.*

Naima, in her twenties, is another example of someone whose mother was told very clumsily by the doctor that her teenage daughter had MRKH and was infertile.

She received a phone call whilst at a Metro station on the way to visit her daughter, who was in hospital for tests:

> Naima: *It was very brief, something like: your daughter has MRKH; she can't have children and she doesn't have a uterus. And there you are, standing in a Metro station.*

A few decades ago, certainly, results of medical investigations which revealed types of intersex/DSD were sometimes shrouded in secrecy. Jacqueline, a woman in her forties, recounts how she found out about her condition after she had undergone medical tests because she had still not started her periods at the age of 16:

> Jacqueline: *So they found out that I had testicular feminisation syndrome. My vagina was, well it was there, but it was minimal. But they didn't tell me, they told my mother. I remember that my mother came to collect me from the hospital and everyone in the ward had had their results, except me. It was all shrouded in a kind of secrecy.*
> Interviewer: *You felt that yourself.*
> Jacqueline: *Yes, I felt that myself; I thought, why aren't they telling me anything? I can still remember that. But they told my mother. I can remember thinking that I was 16, and I said: 'But I've got a right to be told myself; surely I can decide that.' I was already something of a rebellious type, I think. I asked my mother about it as soon as we were in the taxi – my mother sat there with such an expression on her face, I can still see it – 'Mum, what's wrong with me, then?' So then when we got home my mother told me that I wouldn't be able to have children and that I was too narrow down below and that I would never be able to go to bed with boys and that I shouldn't go anywhere near boys. That's an awful lot to take on board.*

It was only years later that Jacqueline was told by a doctor that she had XY chromosomes. Another woman, who is now in her sixties, also heard from a doctor decades later what the real reason was for all the operations she had undergone as a young girl. For both women, this secrecy by doctors had a heavy impact (see also § 3.5 and Box 3.3). As far as we are aware, that kind of secrecy no longer occurs.

Medical advances mean it is increasingly possible to identify certain conditions or anomalies immediately after birth, during pregnancy or during the embryo selection phase. It may be that discovery of an intersex/DSD condition prompts a decision to select a different embryo for fertilisation. Patient organisations sometimes come across couples who are expecting a child and who approach them for advice on whether or not a pregnancy should be terminated because of an intersex/DSD condition. Research shows that the way in which doctors communicate the diagnosis, and which medical specialist advises those concerned, can influence the decision to terminate a pregnancy (Marteau et al. 2002; NNID 2013).[1] Some of the interviewees in this study with intersex/DSD find the idea of terminating a pregnancy very painful: 'It's as if your life isn't worth living' (woman with XY chromosomes). Another woman, Els, who has Turner syndrome, also finds it worrying and painful that when Turner syndrome is discovered during pre-natal screening for chromosomal disorders (e.g. Down's syndrome), those concerned are asked if they would like an abortion. 'Doctors just ask the question, straight out. First they go through the whole list of problems: heart, kidneys, hearing, thyroid, motor deficiencies, spatial awareness: would you like an abortion?' This illustrates that doctors consider this to be a serious condition. But Els knows plenty of examples of women with her condition, including herself, who play a full part in society. 'In reality, they are virtually saying that my life is worthless. And who decides that? I think that's awful.' Termination of pregnancies because of intersex/DSD is an issue that is being raised by advocates (ILGA Europe 2013; NNID 2013).

3.2 The medical side: experience of treatments

The nature and intensity of the experiences that people with intersex/DSD have (had) with medical interventions in their lives varies widely. Without seeking to be complete, we give a few examples here which illustrate the spectrum of experiences.

For some people, the medical complaints are mild and their external appearance does not necessitate medical intervention or monitoring. In boys with Klinefelter syndrome or with XYY chromosomes and in girls with XXX chromosomes, for example, the medical complaints can vary and some will rarely or never come into contact with a doctor, while others experience many problems in their daily functioning and seek help.

With some types of intersex/DSD, those affected face a lifetime of medication, often in the form of hormone therapy. Women are prescribed oestrogen, while men take testosterone, because those affected do not produce these hormones or else they are not properly processed. Hormones can have a radical effect. One man with Klinefelter syndrome explained that taking testosterone took a great deal of getting used to, both for him and his wife. His bodily fat distribution changed, his muscle development increased, he became more aggressive, more 'macho', needed less sleep and became more interested in sex.

Some treatments do not require surgery. Women with certain conditions (such as MRKH and AIS) may have a blind[2] or shallow vagina, which can cause them great psychological difficulties. Insertion of a prosthesis (vagina dilator) can allow them to exert pressure to deepen or expand their vagina. This practice, which has to be carried out on a daily basis over a certain period, can be an alternative to surgery to create a vagina. However, the use of the dilator does not always work and some women give up because it is too painful, too demanding or mentally too difficult.

Surgery on the internal or external sex organs may involve a single operation, but in some conditions several operations are often required because of the complexity, because new problems arise in a new phase of life or because new complications continually arise. The accounts of Barbara (Box 3.1) and Paul (Box 3.2) describe the treatments and operations that were required because of their conditions and how this affected them.

Box 3.1 A lifetime of medical treatments and operations

Barbara, a woman in her forties, has a long history of medical interventions, which continue to this day. At the age of 13 she went to the doctor because her mother thought it was strange that she had not begun menstruating and was not developing breasts. This situation did not change, and at the age of 15 she underwent her first operation. She was told that her ovaries and uterus had developed into a malignant tumour and had to be removed. From that moment onwards, she would have to take hormones. Because her breasts failed to develop, she was fitted with a breast prosthesis at the age of 16. She and her parents were still not told what the underlying diagnosis was. The doctors advised them not to talk to others about it. The result was that Barbara ended up feeling very isolated and built a wall around herself. Some time later her vagina was extended using a portion of her large intestine. An infection developed years later, resulting in severe pain, fever, frequent discharge, feeling constipated and severe cramps. These problems became steadily worse. In recent years she has attended several medical centres and been referred to all kinds of specialists, but no one was able or willing to treat her. Ultimately she ended up seeing a specialist who was able to treat her as a priority, which meant that she could be operated on within four months. She realises that her medical situation is complex, but for her the most frustrating thing has been that different departments were not willing to communicate with each other and the doctors sometimes failed to treat her with respect. Her most important message is that intersex is something natural and should not be seen as an abnormality. She is a woman with X and Y chromosomes, not a girl who should have been born a boy. According to her, it is difficult to be open, because it means talking about 'intimate' parts of the body. This also makes it difficult to let others know that it is sometimes hard to cope with.

In cases where ambiguous external genitals are discovered in babies and young children, surgery is carried out immediately after birth or at a very young age. The surgeons try to bring the appearance of the genitals more into line with the medical norm for a girl or boy. This process is often stressful and distressing for parents, and it can be difficult for them to take decisions about medical interventions. The information they receive from medical professionals, and the way it is conveyed, is important (Streuli et al. 2013). Advocates and human rights organisations are critical of early surgical interventions, and so are some medical professionals,[3] who argue that the physical integrity of the child is compromised if they have no opportunity to give their consent to such irreversible procedures. They also argue that genital operations are not always medically necessary, and are primarily normalising or cosmetic in nature (ILGA 2014; Commissioner for Human Rights of the Council of Europe 2014; NNID 2013; Council of Europe 2013; Méndez, United Nations 2013; Wiesemann et al. 2010).

Box 3.2 Catheter and penis construction

Paul, a man in his thirties, has bladder exstrophy, a condition in which the bladder is not fully closed and is exposed on the outside of the lower abdominal wall. The condition was diagnosed immediately after his birth. Many operations followed, partly because of the complications.

'A leak would develop and I would suddenly pass water from my abdomen; then it was closed up and then one time there was a stone there. There was a leak somewhere and that caused a bladder stone to form. I can't even remember all the operations.'

He was incontinent throughout his youth and had to wear large incontinence pads. At the age of 15, he decided in consultation with his parents that he didn't want to carry on like this. He underwent a major operation, in which a stoma was fitted which was made from his appendix. His bladder was enlarged using a portion of his intestine. 'Yes, that operation. After an operation like that, you have to use catheters to pass water.[...] That was a real breakthrough. And it was great that I suddenly had complete control, including throughout the day.' As a consequence of the bladder exstrophy his penis was very small, and he found that a problem. When he was an adult, he investigated whether it was possible to have his penis enlarged. It was not possible in the Netherlands, so he ended up in Belgium. The surgeons took skin from his arm to use for the reconstruction. It was a difficult time, but he is very pleased with the result: 'For me, it was above all about social freedom – taking a shower at the sports club, using a sauna – and sexual function, to make it large enough for penetration. And of course, there's also the feeling of being 'complete' as a man.' He feels that he is now able to manage his condition very well and that he has got over the social, psychological and sexual difficulties. But he also adds that this is not the case for all people of his age or older.

Removal of the internal sex organs is occasionally carried out because of the small risk of cancer. A number of older interviewees had experienced this. They saw this as a taboo, because they were not involved in the decision to operate; in the past, these operations were often carried out without providing full information to parents and children. They were also not always properly informed by their doctors after the operation and suffered from infertility and osteoporosis later in life as a result of the removal of the internal sex organs. Today, full and proper disclosure to children and parents is the norm among medical specialists.

In addition to the cosmetic and physical outcomes of medical interventions, present-day medical research also increasingly gives overriding priority to 'quality of life', or how patients experience interventions physically, psychosexually, psychosocially and emotionally (Callens 2014; Hughes et al. 2006). Medical and psychological researchers are also trying to gain a better impression of how satisfied people are with medical interventions that they have – or have not – undergone, with allowance being made for the type of intervention, the person's age and the condition. A recent study by Callens et al. (2012a), for example, shows that the majority of women with DSD who undergo operations are reasonably satisfied with their external appearance following surgery, but less satisfied with their sexual functioning compared with other women.

3.3   Self-acceptance of a chronic condition and (sometimes) infertility

It can be difficult to come to terms with having intersex/DSD, especially where the nature of the condition is severe. Most of the interviewees with intersex/DSD have progressed a long way in their process of self-acceptance, but none of them found it easy to come to terms with their condition and its consequences. The process of self-acceptance appears to be focused primarily on coming to terms with having a chronic condition, its physical consequences and the person's self-image as a man or woman. This latter aspect is discussed in the next section.

People with intersex/DSD have to accept that they have a chronic physical condition which may necessitate one or possibly many operations, physical discomfort and limitations, a lifetime of using medication and, in some conditions, permanent monitoring by doctors. Being a patient and having to deal with the condition every day of their lives can be difficult.

Those affected also have to come to terms with the impact on their external appearance. Examples include the appearance of the external genitals in some conditions; a different distribution of bodily fat, breast formation and reduced muscle mass in men with Klinefelter syndrome; lack of underarm hair and body odour in women with some forms of AIS; a less gender-conforming appearance of some women with CAH and persons with PAIS. A consequence of some conditions is that those affected are infertile or become so as a result of operations. Receiving this news often has a great impact on them because it can shatter their picture of the future. Earlier Dutch research has shown that the sadness felt by women with DSD about their infertility sometimes leads to feelings of inferiority compared to others (Callens et al. 2012a). Naima recounts how she heard the news about her infertility in the hospital during her adolescence:

> Naima: *I had just come round from the anaesthetic when I was told. The hospital doctor and the gynaecologist were standing next to my bed and told me: you have MRKH and you won't be able to have children and your vagina isn't, er, complete, well actually you don't have one. They said I had no uterus, effectively. And then that woman agreed with the gynaecologist: yes, but I wouldn't do anything about it, because that's how you were born, so you'll just have to accept it. So there you are, you've just come round [from the anaesthetic] when they tell you all that; you can imagine, it was all a bit strange, to say the least [...]*
> Interviewer: *Can you remember how you felt when you were told?*
> Naima: *I had mixed feelings, really. Something like, oh, I can't have children and I really wanted children. I've always had a certain picture of a future with children and a husband... And then you suddenly realise that you can't have children. So it's more than just not having a vagina. That was the worst thing, really.*

Bart, a man in his thirties, was diagnosed with Klinefelter syndrome when he and his wife were tested because she did not get pregnant. The news that he was probably infertile came as a shock and had an impact on his self-image.

> Interviewer: *You said, 'it was one of the darkest periods in my life'. What was it in particular that was so... Can you say something about that? About the grieving process and what it was that you...*

Bart: *You just start really doubting yourself. I also wanted to leave my wife. I was left feeling: This is my fault. You must now go and do something different with your life, without me, because I can't give you what's normal and what anyone would be able to do. I've never seen her so angry. So okay, that wasn't an option.*

In section 4.2 we look in more depth at infertility and at alternative ways of fulfilling the desire to have a child.

3.4 Self-image as a woman or man

People with intersex/DSD typically see their condition as something that they *have*, not something that they *are*. As far as we are aware, people affected in the Netherlands have little need to be acknowledged as *being* intersex or having an intersex identity. Above all, most of them want to be regarded as normal men and women. People with intersex/DSD almost never have doubts about whether they feel they are male or female. That feeling is generally congruent with their appearance and with whether they were brought up as a boy or a girl.[4]

Intersex/DSD can however have consequences for a person's self-image as a complete or normal man or woman, especially those conditions which are externally visible and where the genitals differ from the medical norm. The sensitivity of matters connected with a person's sex and sexuality also plays a role here. The anxiety about not being regarded as a full man or woman would seem to stem from a feeling that there is doubt about their gender (Alderson et al. 2004). Self-acceptance is important for a positive self-image, but so is knowing that others see you as a full man or woman. The interviews showed that some people with intersex/DSD have had difficulty in seeing themselves in this way. A woman from the focus group who has MRKH said the following about this:

You don't fit the prototype of a woman. For us, of course, things are also a bit different than... You were born without a uterus and without a complete vagina and that's more about the perceptions of other people as to what a woman should look like. And you're not like that, and that's wrong. That's something that's especially common in a man's world, the way people think about that.

Interviewees who are or have been active members of patient associations say that women within their organisations take a great deal of trouble to ensure that they appear well-groomed and feminine. They feel themselves to be women, but sometimes feel the need to emphasise that even more because of their condition. Sjaan, a woman in her sixties, recounts how she felt that she had to confirm her femininity when it became known that she had XY chromosomes. For a long time she was afraid that others could see that. A doctor once made that insinuation, and it affected her deeply.

Interviewer: *You said that your first reaction was: my husband might want to leave me.*
Sjaan: *Yes.*
Interviewer: *Did it also affect your self-image?*
Sjaan: *Yes, of course it did! Really, I just wanted to look as feminine as possible. Because just imagine... And I did imagine that because that one doctor had said it... Yes, of course that does something to you.*

People sometimes need time to accept that their own genitals are different from the average, and may feel lonely during that process of self-acceptance. This also applied for Jacqueline, a woman in her forties with xy chromosomes. For a long time she had the feeling of not being a complete woman and had difficulty in accepting her external appearance.

> One day my father came to my bedroom. I remember that I had a sort of very wide, brown sticking plaster down below, that I was using to cover myself up. I was in front of the mirror. I remember that I had a mirror in my bedroom. A great big one. And I was sitting there sticking the plaster on. When I think back to it now, as an adult, I think: I didn't want that lower part of my body. I didn't want to have anything to do with it. With some kind of duct tape or something [...] One way and another, it's always stayed in my mind: I'm half man. Because of that I could nev-er... I couldn't be happy. I had it in my head that somehow I wasn't right, I was an alien, I was... I came from another planet, I was 'only the lonely'. There was a sort of loneliness in my condition. I didn't have any other people around me who had the same thing.

For Jacqueline, meeting another woman with ais was the affirmation she needed in order to be able to feel she was a *woman*. She recalls the first moment of that meeting:

> We just sort of sat there thinking something like, gosh, you're just a very pretty young woman! She needed that, too, and I think I needed it very badly from her.

There are professional experts who believe that the process of self-acceptance by men war-rants more attention, because they appear to be less organised and generally find it less easy to talk about sensitive subjects. Masculinity is a narrowly defined concept in society, and this becomes all the clearer when men differ from the norm, for example if they de-velop breasts or if the length of their penis is unusual. It is not possible on the basis of this exploratory study to make a good comparison between self-acceptance and self-image in men and women, respectively, with intersex/DSD. The following quotation from a man in his thirties shows that a negative self-image is in any event not the sole preserve of women.

> It has a huge effect on you. It made me feel very different. It sounds very odd, but I genuinely went through a period when I thought: you have boys, you have girls and you have me! You know, a sort of, a kind of in-between feeling, or not really in between, more 'other', you might say.

Various medical procedures are available today in which the body can be 'constructed' to bring it closer to conforming with the medical norm for a man or woman. For Paul, an operation on his penis helped with his self-image. It was stated in the focus group that medical interventions can have a positive effect, but that they are not always the answer to improving someone's self-image as a woman or man. Ultimately, each individual has to go through an internal acceptance process.

> Nicole: *Just creating a new vagina doesn't mean the problem is resolved.*
> Nadine: *It's also about your self-image.*
> Interviewer: *Can you say something more about that?*
> Nicole: *Well, I think that if you have a very negative view of yourself and don't see yourself as a complete woman, just having a new vagina doesn't help. You then have to... It doesn't solve any-thing. On the other hand, if you do see yourself as a complete woman and have a positive attitude*

to life, having a vagina could help. But that's really a very individual thing. It depends how far along you are in your development, in that process.

Callens also argues in her thesis that medical intervention does not always lead to a 'correction of a person's self-image' (2014: 298). She argues that a medical approach should therefore be combined with psychological support.

It is not only the person with intersex/DSD who has to go through a process of self-acceptance: sometimes their parents do, too; they also have to learn to come to terms with loss and anxiety, and sometimes need to adjust their perceptions about sex and sexuality (Alderson et al. 2004; Liao & Boyle 2004b). It is not unheard of for mothers to feel guilty about their child's condition, because they have 'passed on' genetic material. This was the case, for example, with a woman with XY chromosomes:

I was there when they had been watching a medical programme on TV about intersex [special edition of the programme 'Vinger aan de pols' in 2003]. I could see that there was a lot of emotion, and... I find it..., I find it very difficult. My mother felt guilty. And I said: 'Mum, you can't do anything about it.' My mother was the carrier of the gene, though. So yes, ... my mother felt guilty. So I said: 'Mum what on earth could you have done about it? You've just found out that you're the carrier – and then what?'

Medical professionals state that the attitude of parents has a major impact on how children and adolescents view their condition. If the parents have a great deal of difficulty in accepting their child's condition, this often impacts on the child's own acceptance process. Providing support for parents is therefore important (Liao & Boyle 2004b; Wiesemann et al. 2010). In specialist medical centres and patient organisations, both children and parents can receive counselling and support.

3.5   Access to good information and support

For a long time, it was usual practice for medical professionals not to tell patients with types of intersex/DSD about their condition, based on the argument that this would not help their well-being. This confirmed the idea that they had something terrible, which must not be talked about (MacKenzie et al. 2009). This secrecy was often experienced as signifying 'unutterable shame', and its impact generally persists for decades (Liao 2003; Liao & Boyle 2004b: 460). Some argue that this secrecy and the fact that children were never asked for their consent for medical interventions are now the most problematic issues in relation to intersex/DSD (Wiesemann et al. 2010: 673). Interviews with British women with AIS and New Zealand women with intersex show that being given full information, in combination with support, are crucial in enabling them to accept their condition (Alderson et al. 2004; MacKenzie et al. 2009). Box 3.3 recounts the story of a woman for whom lack of good and complete information about her condition affected the course of her life.

Box 3.3 Fighting for information about what you have

Sjaan is aged around 60. For the first 40 years of her life she did not know that she had X Y chromosomes. She underwent operations during that time, but was never informed about her condition. Yet this disorder, as she calls it herself, has had a significant impact on her life. At the age of 16 she went to the G P because she had not started her periods. 'Everything's just a bit small and not fully grown; that'll come in time. So you go back again later; the same story, and I should say that we're almost talking about prehistoric times here.' She has had numerous operations over the years, when the information she was given was not always complete or correct. And she was told that her ovaries had been removed. Later, her uterus was also re-moved. When she was in her twenties, she was told after a medical examination that she was infertile and would never be able to have children. She was not told why.

In her younger years, with financial support from her husband, she gave up her job in order to dedicate herself completely to elite sport. Doctors already knew at the time that she would never be allowed to participate in international competitions because she would be disquali-fied in any examination because of the X Y chromosomes. The prevailing idea at the time was that her muscle development was 'masculine' because of the X Y chromosomes. That view is now outdated.

When she was aged around 40, she suffered severe back complaints and was informed by a doctor on the telephone that she had osteoporosis. In passing, the doctor added that this was not unusual, because she had X Y chromosomes. The endocrinologist explained: 'So then he said: you've got X Y chromosomes in all your cells. You should have been born a boy, but you became a woman. Well, I can tell you, my world fell apart. […] Then I thought: my husband will leave me. What will he want with such a weird person? Category X Y. That's what it was. Your world is turned upside down.' In that period, she wanted to look as feminine as possible.

She saw lots of medical professionals and medical examiners. Her medical records had been locked away in the hospital safe in the past; at least, that's what she had to say when her re-cords were missing. It was only after repeated requests that the employment insurance agency (U W V) was able to have full sight of her records. The U W V ultimately classified her as 'not fit for redeployment', without giving reasons. She herself says that she was deprived of a lot, becau-se others made choices for her without informing her. According to her, the taboo is so great because it is about sex. The worst thing in her view is that she had to fight to obtain infor-mation about what she had. She also had to go to a great deal of effort to get in contact with people like herself, because doctors offered no help. She has now found other women and says that these contacts feel like a warm bath: 'You only need to say one word, and that's enough.'

Today, full disclosure of all information is the guideline for medical professionals who are treating people with intersex/D S D. In the Netherlands, children and adolescents are informed in stages in specialist medical centres, so that the provision of information is matched to their development (Callens 2014). Yet a few of the adult interviewees with intersex/D S D felt that they were still not being given adequate information about their condition and the treatments. The amount of information given by doctors appears to vary widely between the different medical centres. This may be linked to the degree to which centres specialise in the condition in question, but centres also sometimes fail to provide pointers as to where more information can be found. Bart, a young man in his thirties, recalls how he and his wife were told by a medical specialist a few years ago that he had Klinefelter syndrome.

> Bart: *He just came straight out with it and said: 'I see dark clouds hanging over your desire to have children. You have Klinefelter syndrome.' That's what he said, just like that. He then went on to explain how a normal penis ejaculates. How often and how that works. Well, I can tell you, my cognitive overload had already been reached. And so had yours [his wife]. You just sat there sobbing. And the doctor could only look at her [his wife]. It was really awful. We weren't given any information to take away with us.*
>
> His wife: *He mentioned the word Klinefelter once and said something about a chromosome abnormality, and that was it.*
>
> Bart: *So of course we googled it afterwards, because that's what we do.*
>
> *[...]*
>
> His wife: *Or at the very least a folder that you get from the doctor. We weren't even given a folder.*
>
> Bart: *Yes, with answers to basic questions about where you could go for help; this or that contact point; this information on the Internet is right, that information isn't.*
>
> His wife: *Because doctors always say 'you shouldn't go and look on the Internet', but if you aren't given any information, then of course you're going to go and look it up on the Internet.*
>
> *[...]*
>
> Bart: *I also didn't know... I started thinking things like, 'Will I die early, is this ...?'*
>
> His wife: *He didn't want to look on the Internet. It was me who started googling. I didn't even know how to write it. I knew there was something about Klyne (I didn't know how to spell it), and chromosome, I remembered that. And I couldn't remember half of what the doctor had told us. Then I read about it and discovered that life expectancy is normal. It's one chromosome too many. It's this and it's that. So one night when he said to me, in tears, 'Am I going to die early?', I was luckily able to say to him, 'No, you're not going to die early'. Because I'd read that on the Internet. And I think these things are really a big deal; I think the doctor should have told us.*

The persons interviewed with intersex/DSD did not obtain their information only from medical professionals, then. Patient organisations play a role by facilitating the sharing of information about experiences with doctors and different treatments. Medical specialists who see the importance of providing good information also deliver presentations to patient associations. The Internet is a readily accessible source of information, but has the disadvantage that the information available is not always presented well. Professionals point out that people can sometimes be shocked by the potential consequences that are described on the Internet, whereas they do not apply for everyone. It is difficult for lay people to assess the quality of the available information. The fact that sensational stories and images are also placed on the Internet can also prevent people with intersex/DSD from being open about their condition, because they are afraid that others will then find pejorative and misleading information on the Internet.

The Dutch Klinefelter Association (Nederlandse Klinefelter Vereniging) believes there is a need for standardised information (Kalsbeek & Platteel 2012). There is currently no good website available containing detailed information on both the medical and non-medical aspects of various types of intersex/DSD. Information is fragmented across interest groups (particularly the Dutch advocacy group NNID), patient organisations and websites of medical centres.

3.6   Health and well-being

Problems that people experience in relation to their condition can impact on their health and well-being. Research among minorities and vulnerable groups suggests that people's perception of their own health and well-being is an indicator of how a specific group in society is faring compared with the population as a whole. Here, we look at what is known on this subject for people with intersex/DSD.

Medical studies suggest reduced psychosocial and sexual well-being (Johannsen et al. 2006; Callens et al. 2012a; Callens 2014). Whether or not people with various types of intersex/DSD in the Netherlands also score below average on general indicators such as a poorer self-image, depressive symptoms, loneliness, suicide, satisfaction and perceived happiness, is not known. Sanches and Wiegers conclude in a study of young people with CAH that they do have complaints, but that they function in a comparable way to their peers (2010). Yet there are indications that persons with intersex/DSD may have reduced health and well-being.

First, health and thus potentially well-being can be affected by the medical aspects of a condition. People with intersex/DSD may have to cope with reduced functions and capacities (including infertility and comorbidity), permanent medication, problems with their physical appearance, being a patient and all manner of physical complaints. Research among adults with AIS suggests that people with this condition can struggle for the whole of their lives with emotional reactions to their diagnosis, including grief, anger and shame (Slijper et al. 2000).

Second, reduced health and well-being may be related to medical interventions and the way in which the person concerned is treated by medical professionals. Interventions may for example be experienced as traumatic due to poor communication, inappropriate social treatment on the part of medical professionals or disappointing outcomes. A sense of being different, inferior or unhealthy can also be exacerbated by medical treatment (Alderson 2004; Karkazis 2008).

Third, well-being can be affected by actual or anticipated treatment by society. Taboo, silence, embarrassment and feeling that others have difficulty dealing with the condition and show little understanding can lead to a feeling of being different (see § 4.3). Many of those interviewed spoke of loneliness, and this has also been found in other research (Alderson et al. 2004). During puberty and young adulthood, in particular, some interviewees withdrew and felt different, alone and isolated.

There is no overall picture of whether and how these three circumstances (medical condition-specific aspects, medical and social treatment by medical professionals, treatment by others in the social environment) impact on the health and well-being of people with intersex/DSD. There is also currently no information on differences in health and well-being between conditions, method of treatment and sociodemographic characteristics. Research on the need for psychosocial support, its methods and effects is also lacking in the Netherlands.

In international research, psychosocial support and contact with others with the same condition are cited as factors that can contribute to improved well-being (e.g. Alderson et al. 2004; Callens 2014; Hughes et al. 2006; MacKenzie et al. 2009). The persons with intersex/DSD who were interviewed for this study largely confirmed this. Those who had actually received psychosocial support said they had benefited from it. Others sought this support later in life because they had not received it hitherto. According to the individuals and professionals interviewed for this report, however, it is important that psychologists are specialised in intersex/DSD. The professionals interviewed stated that the psychosocial support given in specialist medical centres is today provided as standard by expert psychologists. It is not known whether this is the case everywhere in the Netherlands. Contact with others having the same condition or problems was considered important by all those interviewed for this report. This is discussed in chapter 4, where we look at intersex/DSD in relation to the social environment.

Notes

1 This study was carried out in eight European countries and took place a long time ago. It is not known whether the information and recommendations given concerning pregnancy termination in the Netherlands differ between medical specialists and medical centres.

2 In a blind vagina, there is no birth canal present between the labia.

3 There are some indications that medical professionals sometimes feel under pressure from parents to perform surgery, while this is not necessary from a medical point of view.

4 With some conditions, cross-gender behaviour, or the desire to change sex, occurs more commonly than in the average population, but these are exceptions (Cohen-Kettenis 2010).

4 The social environment: openness, treatment by others and participation

The principal focus in this chapter is on the extent to which having intersex/DSD impacts on the relationship of those affected with their social environment. We investigate whether people with intersex/DSD are open to those around them about their condition and whether they adapt their behaviour. The impact of intersex/DSD on intimate and sexual relationships is also discussed. Some intersex/DSD conditions are accompanied by infertility; we look at the reactions to persons who are unable to have children and at what alternative ways people sometimes find to fulfil the desire to have a child.

The chapter then describes how people with intersex/DSD feel they are treated by others, in other words how others in their social environment react to their condition. We draw a distinction between people in their personal network and professionals in medical practice. We then explore whether intersex/DSD can create impediments to participation in education, work and leisure activities. Finally, we look at the need for contact with others who also have intersex/DSD. Patient organisations exist for some conditions, making it possible to share experiences and information.

As in chapter 3, the different themes and problems are explored on the basis of the experiences of persons with intersex/DSD who were interviewed individually or in a focus group, supplemented with information from a literature search, interviews with professionals and informal conversations. The interviewed persons with intersex/DSD had all received medical treatment and had at some time been in contact with patient organisations; they therefore form a selective group.

4.1 Lack of openness

For people who are aware that they have intersex/DSD, the decision on whether or not to be open towards other people about their condition is one that constantly recurs. One question which may arise is whether it is relevant for those with intersex/DSD to share information about their condition with people other than their partner and doctors. This is after all a highly intimate matter. For many of the interviewees in this study with intersex/DSD, however, not being able to talk completely openly about their condition is a heavy burden. Three women in the focus group spoke about this:

> Woman with XY chromosomes: *I often have the feeling that I can't be completely open and that doesn't feel entirely honest. And I think that friendship is also about honesty. And it's only when you feel that you can tell someone everything that you feel really close to them; but you have to go through that phase earlier.*

Woman with MRKH: *Yes, it's something you have to get over every time. I have it every time I get a new manager at my work. Of course I do a great deal for [patient organisation]. So I sometimes have to leave work, and sometimes I receive phone calls there. And then I have a new manager and I have to explain the whole story again. And then I think: Should I tell him or not? And at a certain point you've taken that step and then the trust is there. But it's a big step before you get to that point.*

Woman with XY chromosomes: *What I've noticed about myself is that once I get to the point where I can be completely honest with friends, in relationships but also in friendships, then I often find that they understand me much better as a result. And then I think something like, why on earth didn't I tell them earlier? People then understand more quickly why you act or react in a certain way.*

These fragments show that keeping the condition secret does not stem from any sense that it is irrelevant. The interviewees want to be open because they have the feeling that they are keeping back something important, but find it difficult to talk about their condition and are therefore unable to be completely themselves. Those interviewees who are active within patient organisations and who could be regarded as a 'vanguard', are also not open all the time and in every situation. Other studies among persons with intersex/DSD reveal the same picture (Alderson et al. 2004; MacKenzie et al. 2009).

Research and interviews with professionals suggest that being open is in fact often a crucial part of the process of self-acceptance and developing a positive self-image (Alderson et al. 2004; MacKenzie et al. 2009). Openness also increases the chance of meeting others with the same condition and of receiving psychosocial support (MacKenzie et al. 2009). Being open is moreover the only way in which people can share their worries or problems regarding their condition with those around them. This applies both for people with intersex/DSD and for their parents. One female interviewee, who was active in a patient association, saw that parents can find it difficult to be open with those close to them about their child's condition:

Woman with XY chromosomes: *It's just about being able to talk freely about something that parents are often very careful about when talking to other people. They often don't even talk to close friends about it.*

Interviewer: *What kind of things, for example?*

Woman with XY chromosomes: *Oh, just being able to talk freely about what's happening with their child and what they are having to deal with and what they miss or what they would like. Or perhaps that they're jealous of the way someone talks about their child if they have a clearly visible disorder... Because you then can't avoid talking about it, because it's visible. So yes, it can be very wide-ranging.*

The study by MacKenzie et al. (2009) shows that people who are open about their intersex/DSD condition find it to be a positive experience, even when it leads to negative reactions or incidents. This begs the question of why people with intersex/DSD are so reticent about being open. People who are aware of their own intersex/DSD condition may be very well aware of the stigma and sensitivities surrounding it. They may feel shame or anxiety about being open because of an anticipated lack of understanding or embarrassment on the part of others. This may be linked to the taboo on talking about deviations from physical sex norms, sensitivities about talking about intimate parts of the body, or sensitivities in relation to infertility (Van Heesch, forthcoming; MacKenzie et al. 2009).

One familiar strategy for avoiding rejection is for people to withdraw and isolate themselves (MacKenzie et al. 2009). Persons who discovered their intersex/DSD condition during adolescence often apply this strategy during that phase of their lives. They become inaccessible to others, both peers and adults. The result is that they may feel lonely, different and not understood and that contact with friends is diminished or lost. For some, this lasts all their lives, while for others it is mainly relevant during their teenage years and early adulthood. Another way of avoiding rejection is to avoid certain situations and locations, so that the openness with others can be 'regulated', as it were. For example, people may avoid saunas and group sports and not allow members of the opposite sex to come too close in a romantic context.

What people tell others, turns out to be an important theme in contact with peers in patient organisations. Many women with XY chromosomes, for example, find it simpler to say they are infertile than to reveal information about their chromosomes. There can be several reasons for infertility and this enables the person concerned to avoid disclosing the cause, namely the intersex/DSD condition. Having a chromosome variation appears to carry a greater taboo (Liao & Boyle 2004b).

Another way of adapting is to try to join in with the behaviour of others. For example, a number of women said that they took tampons with them to secondary school and pretended to have periods, whereas this was not the case. Paul, who had to take incontinence pads to school and change them there, had to try very hard to keep his condition secret: 'Yes, that was awkward. I mean, it felt sneaky and I was in puberty and you feel embarrassed about it, because it's a secret. That has a big impact on you, and it's not nice.'

To what extent the present younger generation find it easier to be open to their peers and to what extent they adapt is not known. Interviewees who are active in patient organisations reported that the way young people deal with this varies widely. It is also not yet known which factors are associated with openness and not adapting.

Box 4.1 We are a forgotten group

Jacqueline is in her late forties and lives in a village with her husband. She has Complete Androgen Insensitivity Syndrome (CAIS). When she was 16 she had to go to the doctor because
she had not begun menstruating. She was not told the diagnosis, but was merely told that she
would not menstruate, was infertile and would not be able to have sex because she had a blind
vagina. Her disorder, as she calls it, was not spoken about in the family. This secrecy was a hor-
rible experience for her. Precisely at the age when her girlfriends were becoming involved with
boys, she was told that she had to stay away from them. She did not even dare to kiss boys
for fear that they would want to go further and would discover her secret. The first year after
learning about her condition was particularly difficult, because she had to come to terms with
the news alone. It was not spoken about at home and she was afraid to take girlfriends into
her confidence. She did not function well and ultimately her school results were so bad that
she had to abandon her studies. She wanted to belong, and therefore always took a tampon
to school with her, even though she had no need for it. When she was in her early twenties,
her rudimentary uterus and ovaries were removed. At least, that was what she thought; later,
she discovered that they were testes and that she had been operated on for something different. Information was withheld and the truth was distorted, and that made her uncertain
and distrustful. Then she learned that she had testicular feminisation syndrome, at the time
also called pseudohermaphroditism. 'That meant nothing to me.' She searched for information, but there was nothing on this topic in the library. Some time later she discovered that the
more common name for her condition was now CAIS. She experienced it as a great burden that
she had to carry this secret alone. She calls it 'invisible misery' and says, 'we are a forgotten
group.' She feels that there is a taboo on talking about her condition, has learned that people
will never ask about it, even when they are aware of her condition, and suffers many physical
and sometimes also mental problems. In her view, the taboo stems from the fact that the
condition is about sexuality. When asked what would help, she says it is important that people
know that there are more possibilities than just boys or girls. Contact with others has helped
her enormously.

4.2 Forming relationships and the desire for children

In many situations, people with intersex/DSD do not need to be open about their condition, but that changes when it comes to intimate and sexual relationships and the desire
to have children. Engaging in sexual contact or embarking on a relationship can be
daunting for them, even if they have a strong desire to do so (Liao & Boyle 2004b). They
may have difficulty with this because of their external appearance or the functioning of
their genitals. This can cause them to avoid sexual contact or may reduce the pleasure of
dating and sexuality (Hughes et al. 2006; Liao & Boyle 2004b). They may feel considerable
embarrassment and fear of rejection (Mackenzie et al. 2009). Patients with DSD often
reach sexual 'milestones' later than the average population and report more psychosexual problems (Callens et al. 2014). Prevailing ideas about what 'normal' or 'desirable'
sexuality is and about what constitutes 'a good partner' mean that they feel vulnerable
and have difficulty embarking on relationships and holding on to a partner.
Research among young Dutch people with CAH showed that those who had experience
with sex and relationships were mostly positive about those experiences, but that there

appeared to be a relatively large group who had no experience (Sanches & Wiegers 2010). A woman with x y chromosomes in the focus group explains why dating can be difficult for girls with her condition:

If you're completely inexperienced and you don't have a vagina, sexual intercourse... you just can't do it. Of course you can engage in sex in all kinds of other ways and you can enjoy it, but you do have to explain to a boyfriend that the most important issue is that you can't do that. And you then start off with a problem from the beginning, because you have to say right from the start: 'listen, we can get it together, but we can't do that.' You have to start off by explaining all that. That's not nice.

Interviews and medical publications show that people with intersex/DSD encounter the reality that 'normal sex' is generally regarded as being limited to heterosexuality with penis vaginal penetration, whereas that form of sex is not always possible for them (Callens 2014; Liao & Boyle 2004b). This can have consequences for their sexual self-image (Van Heesch, forthcoming; Liao & Boyle 2004b). Some young women who were interviewed had kept men at a distance for a long time. Naima, a woman in her twenties with MRKH, talks about this:

Naima: *Somehow or other, there's something that holds me back... And then I think, because I don't have that vagina, er... Yes that's a sort of obstacle. Mentally, I think it's to do with that.*
Interviewer: *That you block it out?*
Naima: *Yes. Then I think, he'll soon want to move on, and then, er...*

For Paul, his bladder exstrophy did not hold him back from dating and getting girlfriends, but his condition did have an impact on his sexual experience. Later he underwent an operation, which helped him feel less inhibited.

It was all fine, really, including with sex, though there was that one girlfriend who said: 'Yes, well I do miss penetration a bit, that it doesn't... Anyway, I miss that a bit.' Of course, you can argue about that, to what extent that should really be a problem.

Persons who are in a relationship when they learn about their condition sometimes fear that they will lose their partner. The condition can undermine their self-image as a full partner. For example, Box 3.3 in chapter 3 contains a description of how the biggest fear of Sjaan, a woman with x y chromosomes, was that her husband would leave her after hearing the news.

For young people, the issue of infertility can play a role in relation to starting or maintaining relationships. Bart, for example, would not have been surprised if he and his wife had separated after it transpired that he had Klinefelter syndrome and was probably infertile (see Box 4.2). Other adult interviewees who are infertile reported that this has or had consequences for their self-image as a partner.

Box 4.2 That's my biggest wish and it's biologically impossible

Bart, a young man in his thirties, and his wife discovered that he has Klinefelter syndrome after a process of medical investigations when his wife did not become pregnant. This syndrome makes the likelihood of pregnancy very small. This marked the start of the most difficult period in his life. 'In the past, I used to talk with friends about "what would be the absolute worst thing if you had to choose", and I always had something along the lines of: if you couldn't have children. So that all sort of came and hit me together when we were told the news. Really feeling that sense of: wow, that's my biggest wish and it's biologically impossible.'

They began looking for alternative ways of fulfilling the desire to have children. Using new medical techniques that were available in Belgium for men with Klinefelter syndrome, an attempt was made to obtain living sperm cells and fertilise an egg cell. This method proved unsuccessful in his case. That came as a heavy blow, partly because the couple were constantly coming across people around them who were pregnant and having children: 'I remember having to go to a party with two pregnant women, for example.' They also saw lots of children in their working environment. They were married and were regularly asked if they planned to have children. It was a sensitive question for Bart. He had always been clear that he wanted children and he had structured his life accordingly. It took him a great deal of effort to abandon that picture of the future. The thought that their parents would never become grandparents was also difficult. In the end they were able to fulfil their desire for children via a donor, and now have a little daughter.

Bart thinks he was fortunate in the way that the Klinefelter syndrome manifests itself in his case, because he has few symptoms compared with other men with the same condition. The biggest impact for Bart was his infertility. He takes testosterone in order to combat osteoporosis and other possible physical consequences. He had to get used to this, and it did slightly change his appearance and his behaviour, but it has had no material consequences for his relationship or his friendships.

Accepting the reality of being infertile not only affects people with intersex/DSD and their partners, but can also have an impact on members of the social network. Some of the interviewees reported that their family and friends often do not know how to react. As a result, there may be little emotional space for dealing with this sadness, or those affected may feel they are not understood. Box 4.3 describes how Els experienced this.

The inability to fulfil the desire to have children, whether or not as a consequence of intersex/DSD, can change the nature of social relationships. An unfulfilled desire for children can be particularly painful or uncomfortable in settings where most adults have children and where that is the norm. Els found that those around her were unable to provide support when she needed it. Barbara began focusing more on her career and Jacqueline sometimes feels lonely and excluded now that her friends are beginning to become grandparents and withdrawing into their family lives. On the other hand, new medical techniques and broader acceptance of non-traditional family types mean that the possibilities for fulfilling a desire for children is increasing for people with a form of intersex/DSD. Examples include donorship, adoption, fostering, artificial insemination, egg cell donation, surrogacy and uterus transplantation[1].

Box 4.3 Oh, now it makes sense. She has solved her problem.

When Els had still not started puberty at the age of 13, her mother had her tested. Chromo-some screening showed that she had Turner syndrome. Her mother found it very difficult to tell her this. Her condition meant that egg cell production stopped at an early age, though the uterus and vagina were present. At the time, the only thing that registered with Els was that she would not be able to have children. That was a major blow for her, because all the girls she knew were already talking about how they would have children later. A number of things now fell into place, including her short stature (1.55 cm), weak short-term memory, reduced spatial awareness and slower social and sexual development than her peers. It bothers Els that doubts are sometimes cast on the intelligence of women like her. She is a university graduate, has a good job and is active as a volunteer. She is sometimes reticent in talking about her condition, because people then go looking on the Internet and find a long list of possible symptoms, whereas no one has all those symptoms together. She has found a way of fulfilling her desire for children: fostering. Women with Turner syndrome can become pregnant via egg cell dona-tion, and she knows women with her condition who have experience of this.

During the period that she had no foster children, she often felt she was not understood, even though people did try. 'I now see how keen everyone is to hear my stories about children. It's also obvious that I greatly enjoy it. I could spend an hour talking about it, no problem. And people also like hearing about it, but there's also a sense of: oh, now it makes sense. Some-times I detect something of an undercurrent in myself, along the lines of, yes, now you're all willing to listen, you're happy to hear this. But where were you when, er... when I was trying to find a way through it.' She finds it worrying that there is less and less space in society to be different.

4.3 Social treatment: ignorance, embarrassment and lack of understanding, but little perceived discrimination

It is difficult to obtain a picture of the extent to which people with intersex/DSD feel subject to negative treatment and discrimination. Thanks to their 'self-management' through limiting their openness, adapting their behaviour and avoiding certain situations and locations, they protect themselves against possible negative reactions. However, it was made clear earlier that not being open and adapting can cause people to feel lonely and different or make them feel that they are not able to be completely themselves. Here we look at how people with intersex/DSD are treated by others when they do decide to be open. How do people react to the news that someone has intersex/DSD? Do those reactions have an impact on social relationships? And do people with intersex/DSD feel that they are discriminated against and not accepted because of their condition?

Reactions concerning intersex/DSD can be related to various aspects: sensitivities surrounding sex or sexuality, physical appearance, physical limitations and infertility. The way in which those close to the person affected react can have a variety of effects: sometimes a friendship becomes closer, so that people with intersex/DSD can be more themselves; sometimes people react with concern, so that those with intersex/DSD find themselves having to reassure them.

Naima: *Because it's not something you can talk about just like that. My tutor was of Surinamese origin. She had the same thing, advising me not to tell other people. Keep it to yourself. In other words, what you now have is really very embarrassing. That's what I thought then, too...*
Interviewer: *Did she explain why you shouldn't tell anyone else?*
Naima: *Yes, she said that people can react oddly to it. In the end, I did explain in the class that I couldn't have children, because I was sometimes absent from school on those days. My tutor, er... everyone said, 'she's never here' and things like that. Something has probably happened... Then my tutor began crying in the class. So she was very concerned about me, but everyone thought I had some kind of serious illness.*

People sometimes ask strange questions out of ignorance or think that the person with intersex/DSD is lesbian or transgender. Some people with intersex/DSD find this hurtful, because it indicates that they are seen as 'different'. In Jacqueline's case, someone once compared her to a well-known transsexual, and she found that very hard to take:

Just imagine, she compared me to a transsexual. So... [...]. Yes, I found it hurtful that my colleague compared me to a transsexual. Yes, it hurt me, I can feel it now. That was something I didn't want. I didn't want to be a transsexual. I have no problem with transsexuals, and that they do their thing. Fine by me. But then I thought: but I'm not like that, am I? I just want to be the same as you. But she couldn't see that, I don't think. That confirmed for me that I was different somehow.

There are also examples of people not talking about the condition or ignoring it.

Jacqueline: *I think it's hard for people around me because it's a sort of charged subject, because it's about sexuality; people don't really understand it very well. They also don't ask about it. That bothers me a bit. Take my parents, for example, who both have [physical condition]. People have no problem asking my mother about that, for example acquaintances that you only see at a birthday celebration; they ask: 'so, [name], how's it going with your [condition] now?' And my mother tells them. But people never ask me that question. It's too dangerous or something, I think.*

The social treatment of other people is characterised mainly by negative reactions stemming from ignorance, embarrassment and lack of understanding. Callens et al. also conclude that ignorance and the taboo surrounding sexuality make communication with others about this subject difficult (2012a).
The impression based on this exploratory study is that the persons interviewed with intersex/DSD do sometimes encounter negative reactions, but do not see this in terms of non-acceptance or discrimination. Total rejection appears to hardly occur, though hiding the condition and adapting behaviour suggest that those affected are afraid that others will not respond in a good way. Whether people with intersex/DSD would encounter much non-acceptance if they were to be open and not adapt their behaviour is impossible to say on the basis of this study. The research literature also makes little mention of acceptance and discrimination in talking about the social problems experienced by people with intersex/DSD. Organisations and researchers focusing on human rights do

relate intersex/DSD to discrimination, but to date have mainly concluded that virtually no information is available on this (see Box 4.4; Agius & Tobler 2012; COC 2013). Interviewees did cite examples of bullying or discrimination that others have experienced, such as girls being accused of being a boy, being excluded and being pelted with sanitary towels. However, it is difficult to assess whether negative reactions and discrimination occur frequently.

We conclude that the interviewees with intersex/DSD do sometimes encounter negative reactions and that people with intersex/DSD adapt their behaviour in order to avoid such reactions. To what extent they systematically encounter non-acceptance and discrimination is however impossible to say. It is possible that persons with intersex/DSD do not see negative reactions in terms of non-acceptance or do not associate with discrimination because the people in their social environment do not wish to be deliberately hurtful or to react negatively. Although negative reactions are sometimes painful, people with intersex/DSD appear to be able to understand them because it is a complex subject. More research would be needed to gain a better insight into how and to what extent negative reactions are related to perceived or actual non-acceptance and discrimination.

The persons interviewed with intersex/DSD are less understanding about negative treatment on the part of medical professionals. With the exception of a man who has always been satisfied with the treatment he has received in specialist medical centres, the interviewees feel frustrated, sad and sometimes angry about how they were or are treated by medical professionals. Some are indignant that doctors know so little about their condition. They talk with a great deal of emotion about poor information, insensitive communication and discourteous treatment.

The need to increase the knowledge about intersex/DSD among medical professionals is broached in virtually every interview. A few specialist medical professionals who were interviewed also raised this point. They reported that medical training devotes virtually no attention to DSD and that anything to do with sexuality is a sensitive area. During the interviews with persons with intersex/DSD, distressing situations were revealed which stem from a lack of knowledge or communication skills on the part of medical professionals. There were examples of doctors who had made incorrect diagnoses, or of persons with intersex/DSD being unnecessarily left in ignorance or 'trapped' in unsuccessful treatment programmes by doctors with too little specialist knowledge. These are rare and sometimes complex conditions, which can be complicated and sometimes also sensational for doctors. Both people with intersex/DSD and medical professionals made mention in the interviews of the importance of specialist medical centres and of medical professionals being aware of them. The medical practice of these specialist centres in terms of information provision and knowledge about these conditions has improved greatly in recent decades. To what extent this is also the case for non-specialist centres is unclear.

Embarrassment and the taboo surrounding sex and sexuality can make lack of understanding or discourteous treatment on the part of medical professionals towards those with intersex/DSD extremely sensitive. Comments or slips of the tongue by medical professionals in relation to someone's sex or gender identity have a severe impact, sometimes leaving people feeling vulnerable and undermined in their physical integrity. Examples include taking photographs without asking permission, or the presence of assistants or extra medical practitioners during consultations or treatment sessions without first seeking consent. The hurt felt by the person with intersex/DSD then does not lie (only) in the nature of the treatments or the content of the consultations, but above all in the way they take place. Jacqueline recounts an old incident that still greatly affects her emotionally:

> Here you can choose between the regional hospital or [specialist medical centre]. So I went. I thought, I need to go to the best doctors, so I went to that [medical centre]. I saw [medical specialist] and he dismissed it as, er, nonsense. He stuck his finger inside me and said: you're wet enough there, and... just nonsense. So he didn't tell me anything at all, and I became unbelievably angry... I was so mad that even in that specialist centre, that, that old... doctor had yet again not told me anything. And then... I can still remember, my husband was with me, and I was, yes, I was really very unreasonable.

It is plausible that medical treatments and social treatment of medical professionals are generally more respectful today. Yet we saw in section 3.2 that medical professionals do still sometimes treat people in a way that is not appreciated.

4.4   Participation in education, work and leisure activities

Based on the information available from earlier research and the interviews in this study, it was not possible to obtain an accurate picture of the extent to which people with intersex/DSD experience frequent impediments to participation in education, work and leisure activities. There is no clear picture of all conditions, and where the consequences for participation are known, the differences between conditions are considerable. Once again, a distinction can be made between condition-specific consequences (e.g. physical limitations or medical treatment) and consequences stemming from self-image or actual or anticipated reactions. The impact on participation is sometimes limited to the period in which persons with intersex/DSD are trying to come to terms with the news about their condition, and their psychosocial well-being is temporarily reduced, possibly influencing their functioning at school or at work. Sometimes, the consequences for participation are longer-lasting and more far-reaching. We give a few examples here of consequences for participation that were cited by those interviewed for this study.

Those interviewees who were aware of their condition at a young age all found that their condition had an impact during their school years. Most of them withdrew for a while, were afraid to be open towards others, adapted their behaviour and felt different or lonely. This resulted in a negative experience of school, reduced social interaction with classmates and anxiety about being bullied. For one person, this led to school results which were so bad that she had to give up her course. It should be borne in mind that all the interviewees were speaking retrospectively and had received little or no support as children or teenagers, nor did they have contact with other young people with the same condition. Today, counselling is offered as standard in specialist medical centres and contact with other young people with the same condition is more easily achieved. It is possible that intersex/DSD has less impact on participation in school and leisure activities for the present generation of children and adolescents, but establishing this would require further research. Sanches and Wiegers come to this conclusion for young Dutch persons with CAH in relation to school and participation in sport (2010). These young people have often been aware of their condition from a very early age, and with good medical treatment and use of medication, the impact on their daily functioning can be limited.

In the small number of conditions involving chromosome variations, it is known that children can experience both social and cognitive limitations and that this can impact on their participation at school and in leisure activities.

Little is also known about the relationship between self-acceptance, self-image, degree of openness, experience of negative treatment and physical impediments in relation to participation in school and leisure activities by young people. What is clear is that the impact of intersex/DSD on a person's youth can sometimes be considerable. For Paul, his bladder exstrophy meant that he had to wear incontinence pads until the age of 15. He found the secrecy and constantly having to hide the pads difficult; it made him uncertain and meant he felt different from other boys. His anxiety about negative reactions proved justified when he took his football trainers into his confidence and told them that he wore incontinence pads:

> But they were just boys themselves, 18 or 19 years old. They weren't the most intelligent boys, either, but one time they just sat there laughing at me. I think I was about 12 or 13 at the time. That's the reason that I stopped playing football.

After undergoing penis construction surgery at the age of 23, he dared to go to the sauna, to take a shower in public and to take part in group sports again.

Intersex/DSD can also have consequences at work, but here again no complete picture is available. Since intersex/DSD is generally not immediately visible for employers, it appears to lead to few problems in obtaining work, though one interviewee did say that her contract was suddenly not renewed just after she 'came out' to her boss. It is unclear whether this happens frequently. Sjaan, an older woman with XY chromosomes, was declared unfit for work because of her condition, whereas she herself felt she was capable of working. This was painful for her, as was the experience of being excluded during her sporting career (see Box 3.3). It is not possible to establish whether situations such as this still occur.

The interviewees cited a number of examples of the way in which the condition can impact on the experience of work or functioning at work, because of physical limitations, new complications, medical interventions or not being open about the condition. For Jacqueline, a woman with XY chromosomes, her condition had a great deal of impact on her experience at work. For a long time she was not open to colleagues about her condition, and when she eventually did take the step, the reactions she received from colleagues were unpleasant, which made her even more reticent about being open. In her case her physical limitations, the taboo surrounding her condition and the lack of acknowledgement by colleagues put her under such mental pressure that the company doctor signed her off work for four months. It was difficult to sustain a working life and she began working part-time. How common it is for people with intersex/DSD to be declared fully or partly unfit for work, or for the working environment or type of work to be adapted because of their condition, is not known.

As regards leisure activities, it is plausible that intersex/DSD has consequences for participation in (group) sports and activities in which the body is visible, such as going to the beach, a sauna or a swimming pool. It emerged clearly from the interviews that this does happen, though it was not possible to establish to what extent.

While the consequences for participation were long-lasting for a few interviewees, not everyone was prevented from participating. Some interviewees experienced problems with participation when they were younger, but at the time of the interview felt no constraints at all. In short, the picture in relation to participation is diffuse and not uniform. It is difficult to establish whether impediments to participating in education, work and leisure activities are mainly incidental or more structural nature. It is also unclear whether there are differences between phases of life and between different conditions, and which factors play a role. Establishing this would require further research.

4.5 Importance of contact with others with intersex/DSD and organisation-building[2]

One thing that all interviewees with intersex/DSD have in common (and this is confirmed in other research), is that meeting other people with a similar condition or comparable problems is important. The feeling of not being the only one with the condition and the ability to share experiences can promote self-acceptance and the ability to deal with the condition. Medical professionals also acknowledge the importance of this. According to Hughes et al., this contact prevents social isolation and increases the sense of being normal (2006). Dutch medical practitioners and psychologists also believe that support groups and patient organisations can offer a platform for sharing concerns and experiences (Callens et al. 2012a). The importance of peer contact emerges consistently in interviews with persons with intersex/DSD (Alderson et al. 2004; MacKenzie et al. 2009). Two interviewees in this exploratory study recounted the following experiences:

And no one thinks it's weird. No one. Whispers: Yes, oh yes, what's that like?... Everyone is understanding. Because of course there's an awful lot going on. You have to imagine: it's a disorder that affects women, with a small vagina, with no vagina, women who have complete AIS and so are insensitive to testosterone, who have to take hormone supplements and go through all that entails. So you can imagine what kind of stories go the rounds at these meetings!

It was truly a... Yes, it was a revelation for me to sit with all those women in a room and find that recognition. Being acknowledged and discovering that I, er..., that I'm not a 'one and only freak show'. That group was a saviour for me. Definitely.

Some people with intersex/DSD only have a need for contact with others with a similar condition during a certain phase, for example when they have just learned about their condition or when they are about to undergo medical treatment. Without exception, however, the interviewees in this study with intersex/DSD had a strong need for contact with others in one or more phases. Some of them did not achieve that contact because doctors did not tell them about the existence of these organisations or because they did not exist during their early years. There was also no Internet in those days where people could look for information. One interviewee who had missed out on contact with others during their youth is thankful to be able to give something back by playing an active part in the patient organisation:

> We now also have parents in the association who don't even have children, who are still pregnant, carrying a child that they know is going to have [condition]. I think it's important that they can see that there is someone there who has a story to tell and who is, let's say, just cool and bright and talks about it freely... That there is hope for them. That they don't need to worry too much that all kinds of things won't be possible.

In the Netherlands, contact with others who have intersex/DSD is currently organised through condition-specific patient organisations. Examples include the Nederlandse Klinefelter Vereniging (Dutch Klinefelter Association), DSDNederland (formerly AISNederland, for XY-DSD), Bijniervereniging NVACP (for CAH), Stichting MRK-Vrouwen (for women with MRKH), Patiëntenvereniging voor Blaasextrophie Nederland (for persons with bladder exstrophy). These organisations organise activities and meetings where information and experiences can be shared. Some adults with intersex/DSD also have contact with Freya, an association for people with fertility problems. Some patient organisations were originally set up as organisations for parents of young patients, and their boards often still consist of parents of children/teenagers with an intersex/DSD condition. There appears to be a growing tendency in some of these organisations for persons with the condition in question to play an active role. As the following quotation makes clear, contact with others with similar experiences can be important for parents:

> What I worry about particularly is the parents. Because I think that providing good support for the [young persons with the condition] begins with the parents. But those parents are so full of emotion and have nowhere to tell their story. I think it's important. Being a parent is already very tough, I see that with all my female friends. And being a parent of a special child, and especially in a society where the scope for being different is reducing, is simply almost impossible.

The organisations where people with intersex/DSD can meet each other have to date been set up mainly as patient organisations. This is related to the historical relationship with the medical profession and the attention for the medical aspects of conditions. External funding is sometimes obtained from the PGO fund (a fund for patient organisations or organisations for people with disabilities, which is sponsored by the Ministry of Health, Welfare and Sport) and sometimes via pharmaceutical companies, again underlining the medical focus of these organisations – though they do not make it a condition that persons actually have to be patients.

The extent to which persons with intersex/DSD conditions have a need for and practical access to contact with others with a similar condition is not known. This information is also not available for subgroups such as children, adolescents, migrants, people with low incomes or people with physical limitations. There are however signs that not everyone with intersex/DSD feels at home in an organisation whose primary focus is or seems to be medical. To date, no organisations have been set up in the Netherlands which specifically target 'non-patients' with intersex/DSD. One exception is the advocacy group NNID, but that does not focus on individuals. OII (Organisation Intersex International), the biggest international organisation for intersex persons, which is established in roughly 20 countries, does not have a branch in the Netherlands. Finally, we would note that cases are known both in the Netherlands and elsewhere of persons with types of intersex/DSD who do not feel or (wish to) behave in a gender conforming way. They may feel more at home within organisations aimed at transgender persons.

4.6   Conclusion

In chapters 3 and 4 we have shown that intersex/DSD not only has physical consequences, but can also impact on personal experiences and social interaction with others. The interviewees with intersex/DSD do not find it easy to be open towards others about their condition. They often think carefully about who they open up to and what they tell them. They also adapt their behaviour or avoid certain situations so that their condition remains hidden. Intersex/DSD can be problematic for relationship formation, a person's self-image as a partner or the ability to fulfil the desire to have children.

There is embarrassment and a fear of negative reactions. Those reactions stem mainly from ignorance, embarrassment and lack of understanding. Most of the persons interviewed with intersex/DSD do not really blame those in their social environment for this and almost none of them seem to associate negative reactions from others with non-acceptance or discrimination. Intersex/DSD can however cause people to feel different, lonely and not understood. This may be related to condition-specific consequences, self-acceptance or actual or expected negative reactions, and the fact that it is not always possible to share concerns among those close to them.

The interviewees with intersex/DSD are less understanding of lack of knowledge and negative treatment on the part of medical professionals. Such experiences are highly sensitive because it is precisely here that they expect to find accurate knowledge and expertise and to be treated with respect.

Contact with others with a similar condition can be very valuable, because those affected can be completely open and can be themselves and are able to share information and experiences. This contact is often organised on a condition-specific basis.

Notes

1   Uterus transplantation is currently not possible in the Netherlands. This treatment could for example offer possibilities for women with MRKH.

2   To prevent any possibility of traceability, the quotes by the interviewees have been completely anonymised as regards the intersex/DSD condition and the interviewees' names.

5   Concluding discussion

The foregoing chapters describe what is known about the social situation of people with intersex/DSD and which themes are important. We would stress once again here that this is an exploratory study. It does not answer the question of how many individuals experience certain problems. In addition to a literature review, our study was based on seven in-depth interviews and a focus group of persons with intersex/DSD, interviews with eight professionals and a number of national and international meetings and symposia. We would note that the persons with intersex/DSD who were interviewed form a selective group and can be regarded as a 'vanguard'. All of them have undergone medical treatment and had contact with patient organisations; in addition, many of them are or have been active in patient organisations. The findings presented here are therefore not representative of persons with intersex/DSD in the Netherlands who are not aware of their own condition or who have never undergone medical treatment.
It became clear during our study that a proportion of the problems or impediments experienced by the interviewees with intersex/DSD are medical in nature and are related to (chronic) physical limitations, including infertility in certain conditions. There are also problems or impediments that have consequences for the social situation. In this exploratory study we attempted to identify all these impediments.
In this concluding discussion, we seek to answer the two research questions formulated in chapter 1, and put forward suggestions for the possible development of policy based on the findings in this exploratory study.

The most accurate estimate that we are able to make at this point in time is that the prevalence of intersex/DSD in the Netherlands is approximately 0.5%, equivalent to just over 80,000 people. It is not known how many people are aware of their condition and have received medical treatment for it at some point.
Summarising, it is apparent on the basis of this study that there are wide differences between intersex/DSD conditions and their consequences for people's daily lives. There are also substantial differences in the experiences of younger and older generations. Developments have not stood still in the medical world, and practices such as secrecy and deliberately providing false information would appear to have died out.
Virtually all the persons interviewed with intersex/DSD have encountered problems in their social situation. They talk about being 'different', feeling lonely and experiencing shame and embarrassment. In most cases – either in the beginning or throughout their lives – they have difficulty in accepting their condition and experience problems in relation to their health and well-being. If the condition has consequences for their external appearance, this can have an impact on their self-image as a man or woman. In addition, intersex/DSD often has an impact on relationship formation and on the ability to have a positive experience of sexuality. This may be due to fear of rejection of their appearance, their self-image as a man or woman, limited sexual abilities and their own perception and/or being informed that they are infertile (in some conditions). Infertility

generally leads to personal sadness, and that is not something that can always be easily shared with others.

Most of those interviewed find it difficult to be open about their condition with those around them. They sometimes adapt so that their condition remains hidden. There is also fear of negative treatment by others, mostly stemming from ignorance, embarrassment and/or lack of understanding. People with intersex/DSD encounter this both in their personal environment and in their contacts with the medical profession. Finally, this study describes how intersex/DSD can have consequences for participation in education, work and leisure activities, though no complete picture can be given of this. What is clear is that physical limitations and reduced psychosocial well-being can mean that people sometimes function less well or withdraw, and this can have an impact on their social relationships with fellow pupils and colleagues, and therefore on their participation. Puberty, in particular, appears to be a vulnerable phase of life, as this is a period when sexuality, intimacy and external appearance are extremely sensitive. The consequences for a person's self-image, well-being, connection with peers and impediments to participation at school and leisure activities can therefore be exacerbated during this period.

5.1 Perceptions and visibility

It is clear that persons with intersex/DSD encounter a great deal of ignorance in society in relation to their condition. Among the prejudices encountered by the persons interviewed in this study are the following: persons with intersex/DSD are something 'between a man and a woman'; they are not complete men or women; they are inferior or not normal; they are gay/lesbian; they are freaks; and they all have ambiguous genitals. Problems experienced by people with intersex/DSD in their social situation and which are not condition-specific appear to be closely linked to these perceptions, which in turn are linked to sex normativity and sensitivities in relation to sex, sexuality and infertility. In the case of infertility, expectations and the social norm of having children also play a role.

By 'sex normativity' we mean that a person's sex is seen as a dichotomy, with just two possibilities, 'male' or 'female' as the unquestioned norm. Everything that falls outside this twofold division is regarded as differing from the norm. However, intersex/DSD shows that, medically speaking, a person's sex is not an absolute dichotomy. Parallels can be drawn here with heteronormativity, in which heterosexuality is regarded as the norm and as natural. Non-heterosexuality deviates from this norm, so that gay/lesbian and bisexual people become involved in processes of coming out, feeling different and fear of stigmatisation.

The literature review and interviews with non-medical professionals revealed that advocates and social science researchers are questioning this dichotomy in relation to a person's sex, and therefore sex normativity. It must be said that the persons with intersex/DSD interviewed in this study did not themselves express this need very strongly. In most cases they have no desire to abandon the dichotomy, but feel unambiguously a man or a woman and wish to be seen as normal men or women.

The fact that the phenomenon of intersex/DSD is relatively invisible in society is linked not only to ignorance but also to the fact that persons with intersex/DSD often wish to be seen as normal people and barely identify or present themselves as a group. Moreover, they tend not to be very open about their condition because of the sensitivities described above.

There is some question as to whether more visibility is the solution. People with intersex/DSD, patient organisations and interest groups react differently to this question. Interest groups and persons active in a number of patient organisations argue that greater awareness and visibility of intersex/DSD in society will have an emancipatory effect, due to altered perceptions and greater knowledge of intersex/DSD, and consequently a reduction in the sensitivities surrounding the conditions. They argue that people would become more accustomed to the existence of intersex/DSD. Raising awareness and expanding the norms regarding men, women and sexuality would also help change perceptions surrounding intersex/DSD. This could enable people with intersex/DSD to be open about their condition without embarrassment, anxiety or reticence.

Others we spoke to in the context of this study expressed fears that more visibility could lead to more stigmatisation. Most of the interviewed persons with intersex/DSD are reluctant to appear very visibly or recognisably in the media. Members of patient organisations say that requests for interviews or media attention are often difficult to grant because few people within the organisation are willing to put themselves forward. Some are afraid of insensitive or sensational attention in the media, fearing that the heightened visibility will lead to more stigmatisation. The interviews with persons with intersex/DSD also revealed scepticism about more visibility of and familiarity with intersex/DSD; they have accepted that there are taboos and sensitivities. This can in fact be part of a coping strategy.

An additional dilemma is that greater visibility in society could lead to people with intersex/DSD being seen as a distinct group, whereas most of them do not wish to be seen in that way. Those interviewed do not see their intersex/DSD as an identity. They have a condition, but generally regard themselves as a man or a woman. It is therefore questionable whether people with intersex/DSD would benefit from more visibility if this also meant that demarcation lines between themselves and people without intersex/DSD were drawn more sharply.

In short, it is difficult to assess what the effect of greater visibility would be. When considering the question of whether more visibility would be beneficial or harmful, it is in any event important to draw a distinction between the individual and societal level, and between the fact that persons with intersex/DSD do not feel part of a separate group, but as a group of individuals do encounter taboos and sensitivities. Whether more openness and visibility would have a positive or negative effect on well-being, self-image, acceptance and participation can also not be established on the basis of the available information. It is plausible that more visibility of and knowledge about intersex/DSD could have a positive effect in the long term if successful efforts are made to reduce taboos and sensitivities, but this approach also carries risks such as negative treatment and stigmatisation.

5.2 Medical practice: need for more knowledge, sensitivity and views on sex

Three issues in relation to medical practice emerged from this exploratory study. First, the interviews revealed that the knowledge about intersex/DSD among medical professionals is generally deficient. Medical specialists also believe that too little attention is devoted to intersex/DSD, sex and sexuality in medical training programmes. In the literature, Sanches and Wiegers (2010), for example, conclude on the basis of their research among children and teenagers with CAH that improvement is possible and necessary in the knowledge of caregivers. As with the care provided to transgender persons, this demands expertise. The importance of specialist medical centres in the Netherlands is broadly acknowledged by all concerned. Improving the knowledge of medical professionals in general is also necessary, because most persons with intersex/DSD initially come into contact with non-specialist doctors and care providers.

Second, almost all interviewees had encountered a lack of sensitivity on the part of medical professionals. Intersex/DSD is linked to a person's sex and sometimes their external appearance, infertility and sexuality. This is an extremely sensitive area for many people. They expect doctors to treat them with respect, understanding and empathy, and when this does not happen they can find this very difficult to deal with. It is plausible that the attitudes of medical professionals today have improved compared with past decades: some medical interventions are no longer carried out; secrecy is no longer the norm; and there is more attention today for the ethical aspects of medical interventions and social treatment (for medical ethics aspects see also Appendix C). Nonetheless, recent examples of insensitivity still exist. It should also be noted that this may not apply only for intersex/DSD, but might characterise the medical profession as a whole.

Third, non-medical researchers and advocates point out that the drive to raise the knowledge of medical professionals should not only relate to medical knowledge and social treatment, but also to their views on sex. Medical professionals regard intersex/DSD as a medical matter, which means that those with intersex/DSD are regarded as people with a medical problem who require medical treatment. However, a focus on medical intervention in the case of ambiguous sex organs sustains the thinking about a person's sex as a strict dichotomy, both in the medical world and in society (e.g. Karkazis 2008; Liao & Boyle 2004a; Wiesemann et al. 2010; Van Heesch 2009). At individual level, people could benefit from surgery which gives them the appearance of what a woman or man should look like. According to Liao and Boyle, however, more scope for and tolerance of intersex/DSD will only be achieved when perceptions change as to what constitutes 'normal' with regard to a person's sex (2004a).[1] What is regarded as 'normal' in medical circles is based not only on biological criteria, but also on social criteria, neither of which are fixed. Wiesemann et al. draw a comparison with homosexuality, which was at one time regarded as a biological and psychological abnormality, whereas this is no longer the case today. Interest groups and human rights organisations believe that the aim should be demedicalisation and that the condition should only be treated when this is medically necessary. Those interviewed with intersex/DSD cited examples that could be linked to this. An active board member of a patient organisation is trying to draw the attention of medical professionals to the fact that 'not everything that is different is a problem. And not everything that is a problem requires a medical solution'. The indigna-tion expressed by some interviewees about prenatal diagnosis leading to termination of pregnancies or other forms of embryo selection (see § 3.1) is experienced by some as disproportionate medical intervention.

5.3 Gaps in knowledge regarding social situation

Generally speaking, it was possible to establish in this exploratory study that certain problems occur, but not to what extent. It was also not possible to draw a clear distinc-tion between subgroups. At the level of social situation, for example, little is known about differences between conditions or about which groups are especially vulnerable. The knowledge of differences based on sociodemographic characteristics such as gen-der, ethnicity and age is limited. For example, are there differences between men and women with intersex/DSD or are the differences mainly condition-specific? Some medi-cal professionals argue that intersex/DSD is a very sensitive matter in certain migrant groups, especially if it is accompanied by infertility. These individuals appear not to seek or find contact with patient organisations. Whether they experience particular problems in their social situation, could be explored, though these persons will not be easy to reach for research. Developments in medical practice mean that older generations have encountered specific problems in their medical and social treatment which are probably experienced less by today's younger generations. The social situation of younger genera-tions may also be expected to be more favourable, though this cannot be stated with certainty. It would also be interesting to examine whether there are differences between people who are receiving or have received medical treatment and who are in contact with others with the same condition, and persons with types of intersex/DSD who do not have this experience.

There are some indicators that are relevant for a person's social situation for which it was not possible to obtain a clear picture based on the interviews and literature review. For example, we can conclude that there are signs that people with intersex/DSD encounter negative reactions and adapt in order to avoid such reactions, but to what extent there is evidence of non-acceptance and discrimination remains unclear. There are also exam-ples of impediments to participation, but the picture is diffuse and incomplete. Do peo-ple with intersex/DSD less often have work? Do they drop out more frequently? Do they underperform or make other choices in terms of participation opportunities as a result of their condition or anticipated negative reactions?

Some of these questions could be answered better using large-scale quantitative research on the social situation of people with intersex/DSD. In Appendix A, we describe how such a study might be designed, what its focus might be and what the associ-ated conditions and challenges are. It also discusses the fact that, while human rights organisations and medical professionals acknowledge the need for research on the non-medical aspects of intersex/DSD, no large-scale quantitative research has to date been carried out from a non-patient-based perspective. The greatest challenges lie in classify-ing the target group and obtaining a non-selective study population of reasonable size. The target group is highly diverse and not everyone will be able or willing to participate, for example people who are themselves not aware of their condition, people who have little interest in participating (for example because they have never had any negative experiences) or people for whom it is too painful or confrontational, as well as people who have little inclination to affiliate themselves with intersex/DSD or with their own condition.

We explored the practical opportunities for carrying out large-scale quantitative research. A combination of recruitment methods would be the best option if new survey research were to be set up. The cooperation of medical centres is a crucial factor here. One advantage of this kind of research is that extensive attention could be given to aspects relating to the social situation, but such research is expensive and time-consuming. We also explored how analyses of themes related to the social situation could be carried out within current and existing patient-oriented research. These options are cheaper and less extensive. One drawback of these options is that outcomes in relation to the social situation would be more limited and that recruitment is carried out among patients. As a result the respondent group, and possibly the outcomes as well, are likely to be more selective.

Children, adolescents and their parents were largely left out of consideration in this exploratory study. This is a potentially vulnerable group about which information is scarce (COC 2013). Exploratory qualitative research could provide more insight here. Finally, we would add the caveat that a number of medical ethics issues were addressed only summarily because they fell outside the scope of this exploratory study, while these issues may warrant more attention.

5.4 Pointers for policy development

An exploratory study of intersex/DSD from an emancipation perspective, as announced by the Dutch minister with responsibility for emancipation policy, is fairly innovative. This question, and any policy proposals from this perspective, is related to the scope and view of emancipation. We do not adopt a standpoint on this question. It was also not possible on the basis of this exploratory study to establish the degree and severity of problems occurring in the total group of people with intersex/DSD, whereas that information could be important in determining to what extent there is an emancipation issue here. In this section we look at pointers for policy development based on the findings on relevant themes, potential problems and gaps in knowledge which have been highlighted in this study.

With a view to possible policy development, we would first note that there are sensitivities surrounding the development of specific policy for 'the group' of people with intersex/DSD. The persons interviewed in this study do not see themselves as a separate group and do not wish to be seen as such. There is a high degree of heterogeneity; people with intersex/DSD include persons who have received medical treatment throughout their lives and who have found it to be a major impediment in their personal, relational and social lives, but also people who are not aware that they have intersex/DSD or who experience virtually no consequences as a result of having it. The absence of a clear group or community makes developing specific policy difficult.

Policy development in relation to problems in the social situation could target the impediments that stem from normativity, sensitivities and perceptions in relation to sex, gender and sexuality. There are points of contact here with two groups that are currently the focus of the national emancipation policy, namely women's emancipation and LGBT (lesbian, gay, bisexual and transgender) emancipation. Correspondences with the situation of LGBT persons include difficulty in being open, low self-acceptance, (fear of) stigmatisation in society and little positive experience of sexuality.

Given the correspondences in relation to social problems, the addition of the letter 'I' to LGBT is becoming increasingly common internationally in the focus on observing and promoting human rights (e.g. ILGA Europe 2013; Council of Europe 2013) and in the foreign policy of the European Union (2013).[2] This also applies for the implementation of foreign policy in the Netherlands (BuZa 2013). Adding an 'I' to LGBT is understandable in the sense that all these groups experience problems because of perceptions in relation to sex, sexuality and gender.

However, the issue highlighted above, namely that persons with intersex/DSD do not see themselves as a distinct group, would also have to be taken into account in deciding whether adding it to LGBT emancipation is the best means of embedding intersex/DSD in Dutch emancipation policy. Most persons with intersex/DSD who were interviewed in this study say they prefer to distance themselves from LGBT persons because they are often confused with them, whereas for them they are totally different groups. The association with sexual orientation or gender identity is sometimes experienced as unfortunate or disagreeable, because these aspects are not central to the problems of people with intersex/DSD. Although the particular condition that people with intersex/DSD have can influence their ability to have a positive experience of sexuality and relationships, in virtually all conditions there is no direct relationship with sexual orientation.

The interviewees with intersex/DSD feel very far removed from transgender persons. This applies especially where people with intersex/DSD feel strongly that they are a man or woman and behave in a gender-conforming way. Those interviewees who stress that they are ordinary men or women will feel very disconnected from an expansion of LGBT to LGBTI. Persons with intersex/DSD whose gender identity or gender expression is more ambiguous may perhaps see more scope for an affiliation with transgenders, but this did not become clear from the interviews conducted in this study.

Regardless of whether people with intersex/DSD and transgenders find some affinity with each other, we would point out a number of clear differences here. Difficult situations because a person's legal identity does not correspond with their external appearance generally do not arise in people with intersex/DSD.[3] There is also no question of a shared identity or community among persons with intersex/DSD. Different medical ethics issues are relevant for people with intersex/DSD, such as prenatal diagnosis and medical intervention at an early age without the consent of the person concerned. Moreover, persons with intersex/DSD can suffer greatly from the widespread ignorance about intersex/DSD in society and among non-specialist medical professionals, whereas transgenders have become much more visible in recent years and receive more attention.

Finally, we would note that some of the problems identified in the social situation of people with intersex/DSD are linked to medical practice or are condition-specific. There are correspondences here with groups wishing to draw attention to comparable medical ethics issues or medical practices. In the case of intersex/DSD, particular attention is moreover needed for the sensitivities surrounding sex and sexuality and the sometimes complex medical picture of certain conditions.

Below we set out a number of pointers for themes where it is already clear that problems occur and where policy attention could make a positive contribution.

Better picture of social situation via research

As became clear in the previous section, there is no complete picture of experienced non-acceptance, discrimination and impediments to participation. There are other relevant themes where it is also not possible to establish the extent to which problems occur and which persons with intersex/DSD are at risk of experiencing problems in their social situation. If there is interest at policy level in the position of persons with intersex/DSD in society, quantitative research could contribute to the knowledge and assessment of how problematic the consequences of intersex/DSD are for their social situation. In the case of homosexuality, bisexuality and transgender persons, quantitative research has identified what the frequently occurring problems and impediments are, and this has given direction to government policy (Keuzenkamp et al. 2006; Keuzenkamp 2010; Keuzenkamp 2012).

Little is known about children, adolescents and parents in relation to intersex/DSD, and their social situation also warrants attention.

Promoting resilience and organisation-building

Section 4.5 describes how sharing experiences and information with others with similar conditions or problems can help increase the resilience of people with intersex/DSD. In the Netherlands, this takes place mainly within condition-specific patient organisations which focus on persons with intersex/DSD and those close to them (e.g. parents). They seek to function as a source of good, understandable information. They also aim to build relationships with and improve medical practice. It is not clear to what extent children and adults with intersex/DSD are made aware of the existence of these organisations, have access to them in practice and feel at home in them.

During a number of meetings organised last year by the Dutch advocacy group NNID (Nederlands Netwerk Intersekse/DSD/Netherlands Network Intersex/DSD), representatives of patient organisations explored what correspondences, shared interests and problems and differences there are between the various conditions. Correspondences include the issues surrounding the ability to be completely oneself and open; the need for medical professionals to provide good information and have sufficient knowledge; problems in entering into relationships and fulfilling a desire for children; prenatal or postnatal diagnosis (see § 3.1). Whether people are troubled by a low self-image as a man or woman, and which medical ethics issues are relevant, varies enormously depending on the condition. There are also organisations which do not fully embrace the fact that they are classified as intersex/DSD according to the generally accepted definitions. Sharing of experiences between patient organisations has been found to be useful and there appears to be a desire to continue this. At the same time, there is a reticence to operate as a single organisation, because the present patient organisations are so diverse in their nature and design.

One practical problem is that some patient organisations have difficulties with staffing and funding. This is because the groups they target are small and potential volunteers may be impeded by medical problems or the taboo that surrounds their condition. Other patient organisations have the problem that the condition they represent is not recognised as an independent condition by the PGO fund (fund for patient organisations or organisations for people with disabilities, which is attached to the Ministry of Health, Welfare and Sport), while collaboration with a patient organisation that is not related to intersex/DSD would be unthinkable for these organisations because of the sensitivities and taboos involved. This can jeopardise the diversity and extent of the activities available for promoting resilience.

Unlike in other countries, there is no organisation in the Netherlands for people who identify themselves as intersex and who reject a medical perspective. It was not possible to establish in this study whether these persons, like people with intersex/DSD who are exploring or questioning the boundaries of sex and gender conformity, make a good connection with patient organisations or other organisations (e.g. transgender organisations). It is also unknown how large this group is in the Netherlands. Finally, we would note that there is an absence of activists willing to engage in a fierce public debate about human rights issues, whereas there are examples of this in other countries.

More knowledge and sensitivity on the part of medical professionals

The perceived need to increase the knowledge, awareness and sensitivity of medical professionals has already been mentioned in this chapter. Efforts to increase their knowledge could focus not only on the medical aspects of intersex/DSD, but also on the social aspects. For non-specialist medical professionals, this would entail knowledge about intersex/DSD, sex and sexuality, and the existence of specialist medical centres. Imparting this knowledge could form part of basic medical training and care training programmes. Promoting sensitivity in the provision of information and treating patients with respect by all medical professionals could also contribute to a positive experience of medical treatments. That also applies for attention for questions of medical ethics (see Appendix C), including views and norms in relation to sex and gender.

Focus on acknowledgement, visibility and awareness-raising

It emerged clearly from the exploratory study that persons with intersex/DSD have a need for acknowledgement, both medically and socially. The widespread ignorance about intersex/DSD in society, combined with the taboo and sensitivities surrounding it, can make it difficult for people with intersex/DSD to feel that they belong and are complete. Taking part in an interview evoked strong sense of acknowledgement in some interviewees, indicating that this is something they generally do not feel.

The study also showed that the relationship between acknowledgement and visibility is a complex one. There are differing views on the question of whether more visibility is harmful or beneficial (§ 5.2). More visibility of and knowledge about intersex/DSD could contribute to reducing taboos and sensitivities, but could also mean that intersex/DSD becomes classed as a new group or category; as indicated above, this does not appear to be the wish of people with intersex/DSD. Moreover, not everyone is familiar with the terms 'intersex' and 'DSD', and not everyone wishes to associate themselves with them. It is therefore unclear whether specific policy attention aimed at raising visibility and awareness and changing perceptions of intersex/DSD in society would have a positive or negative effect in improving the social situation of people with intersex/DSD. A strong focus on intersex/DSD is not the most obvious approach in seeking to increase visibility and awareness among the general public. In the societal context, an approach based on normativity and perceptions in relation to sex, gender and sexuality, and allowing space for acceptance of diversity in appearance, could be a better option. This is related to more general attempts to raise awareness about stereotypes and implicit norms in relation to sex, gender and sexuality in society, as well as fostering understanding of the impact on people who do not comply with those norms. When it comes to medical professionals, it is vital to talk specifically about DSD or intersex.

Specific examples include information provision and open discussion of sensitivities and normativity in relation to sex, sexuality, gender and infertility. In the first instance, this could be aimed at professionals, and also at a broader public. Positive media attention was also cited in the interviews. Textbooks on biology could mention that there are more possibilities than 'male' and 'female'; the information provided in biology textbooks should in any event be correct. A website providing accurate information is important for persons with intersex/DSD and those around them. This could contribute to a positive self-image and reduce the fear of negative reactions, because it would be possible to refer people to a website that does not contain pejorative information.

Promoting knowledge-sharing

In the context of this exploratory study, contact was made with various medical professionals, patient organisations and advocates. It was notable in the discussions, interviews and meetings that the distance between these organisations and parties is substantial. The use of language, type of knowledge, themes that are important for them and the way in which intersex/DSD is viewed from different perspectives, varies widely. A few examples will show how fragmented the knowledge is and how much the perspective on intersex/DSD can differ between organisations and parties. Non-specialist advocates are strongly focused on human rights and issues of medical ethics, while their knowledge about the medical aspects of intersex/DSD is sometimes limited. Medical professionals and (some) patient organisations often have little awareness of international developments in relation to the promotion of human rights and non-medical advocacy. For medical professionals and (some) patient organisations, terms such as emancipation, discrimination and human rights are sensitive or remote concepts, whereas this is the primary terminology used by advocates. At this point in time, there is no established platform where knowledge and views are shared between these organisations and parties. As a result, organisations and professionals are sometimes not aware of developments in various areas in relation to intersex/DSD which could have an impact in improving the social situation of people with this condition.

With support from the Ministry of Education, Culture and Science, the NNID has organised a number of meetings in the past year at which representatives of patient organisations and knowledge institutes, a number of researchers and medical professionals came together to share knowledge and experiences. This was felt to be useful and was appreciated. It also emerged during these meetings that the distance between the different organisations and parties was sometimes considerable, though there does seem to be a desire to share knowledge and experiences.

Since its foundation, the NNID has served as a point of contact for the government. It disseminates information on national and international developments and places societal problems in relation to intersex/DSD on the agenda, including problems affecting people's social situation and questions relating to medical ethics and other human rights issues. In other groups which are the target of emancipation policy, advocates contribute in a comparable way to promoting knowledge, drawing attention to problems and initiating discussions. Information provision and knowledge-sharing from different perspectives is important in promoting the social situation of people with intersex/DSD.[4]

Conclusion

During this exploratory study, many interviewees were grateful for the acknowledgement and attention their stories received, and for the lifting of the taboo surrounding the problems they encountered as a result of their intersex/DSD condition. During the past year, the international attention for these people has grown. Developments have also not stood still in the Netherlands, as illustrated by the announcement of an exploration of the issues by the minister, the founding of the NNID and attention in the medical profession through research and publications. This is clearly a field where things are developing.

Notes

1 To indicate that deviation from medical standards is normative, Morland states that a small penis is perceived as more problematic than a large penis (2004: 449).

2 Human rights issues fall outside the scope of this study. For a brief look at the human rights perspective, see Appendix C.

3 There are exceptions, though it is not known to what extent this occurs among people with intersex/DSD.

4 Advocacy is no longer funded by the PGO fund since 2014, and it is therefore uncertain to what extent patient organisations will be able to (continue) fulfilling this role. The extent to which patient organisations would like to focus on this activity varies.

Living with intersex/DSD

A.1   Usefulness and need

Research from a non-medical perspective on the experiences of people with intersex/ DSD is still scarce. The Council of Europe, a study carried out at the request of the European Union, and interest organisations point to the lack of solid, research-based knowledge about people with intersex/DSD (e.g. COC 2013; NNID 2013; Agius & Tobler 2012; Council of Europe 2013). Medical professionals also conclude in a broadly supported publication that the clinical world is strongly focused on gender and genital appearance, whereas stigma and experiences within and outside the medical setting can be much more pertinent questions for those concerned (Hughes et al. 2006). Callens et al., in the first large-scale Dutch patient-based study of people with intersex/DSD, also observe that more research is needed into the consequences of social stigmatisation (2012a). In short, the calls for research from a non-medical perspective on the experiences of people with intersex/DSD appear to be growing.

To date, no such large-scale quantitative research has ever been carried out, either in the Netherlands or internationally. Information on the social situation of people with intersex/DSD in the Netherlands is scarce and derives mainly from small-scale, sometimes condition-specific qualitative studies (e.g. Sanches & Wiegers 2010 on children with CAH; Kalsbeek & Platteel 2012 on patients with Klinefelter syndrome; Van Heesch on the relationship between life stories and medical narratives in relation to intersex, forthcoming as a dissertation in mid-2014). In patient-based quantitative research, questionnaires do often contain a number of questions on non-medical topics, such as self-image and well-being, but discussions with the researchers carrying out these studies show that reporting on social situation does not always have a high priority, though they recognise its importance.[1]

This exploratory study identifies the areas in which people with intersex/DSD may experience problems in their social situation. It looks at their personal experiences and the relationship with their social environment. On a number of aspects, it establishes that problems occur, for example in relation to self-acceptance and self-image, well-being, lack of openness, negative treatment and inadequate provision of information. The study also finds that there are some themes about which little is known, for example how and to what extent non-acceptance, discrimination and impediments to participation occur (though this is due in part to the small and selective group of persons with intersex/DSD who were interviewed). It is also unclear which factors are associated with well-being and with participation. It was not possible to establish on the basis of this exploratory study to what extent specific problems occur. It may be that the limited size of the study meant that certain problems were not recognised. It was also not possible in this study to look at differences between groups, for example between generations and between conditions. Large-scale survey research could provide more insight into these differences.

Here we explore the conditions and practical options for survey research to study the social situation of people with intersex/DSD. First, in section A.2, we suggest some of the themes that would need to be covered in such a study. In section A.3 we then look at methodological issues relating to the classification of the target group, attention for subgroups, possible recruitment methods, reaching potential respondents and nonresponse. Three practical options for conducting survey research are described in section A.4, as well as the advantages and drawbacks of each. Finally, in section A.5, we put forward suggestions for topics where, in addition to quantitative research among the target group, more knowledge is needed about intersex/DSD.

A.2   Themes for research in the target group

With a view to the possibility of carrying out quantitative research on people's social situation, in this section we summarise the findings from the exploratory study concerning problematic aspects and themes about which little is known but which could be relevant. We supplement this with indicators that are commonly used in research among other minorities, such as LGBT persons (lesbian, gay, bisexual and transgender persons) and migrants. Like people with intersex/DSD, members of the latter group may also have to deal with negative perceptions, non-acceptance, discrimination and impediments to participation.

Relevant themes in relation to people's personal experiences of intersex/DSD are their 'discovery' of their condition and their own reaction to it. Some conditions also involve coming to terms with aspects of external appearance and infertility, and can also impact on the person's self-image as a man or woman. How do persons who are not open towards others about their condition cope with carrying a secret, and which coping strategies do they use? Have people received good information and support? How do they perceive their own health and well-being, and which factors are associated with this? Earlier research among LGBT persons found that psychosocial and physical well-being, loneliness, perceived health and happiness are good indicators for gaining an impression of these aspects (e.g. Keuzenkamp 2012; Keuzenkamp et al. 2012; Kuyper 2013).

Whether or not people with an intersex/DSD condition are open to people in their social environment proved to be an important theme, as did adapting their behaviour or avoiding certain situations. It also became clear that negative treatment and negative incidents do occur because of the condition. There is no clear overview of the form, degree and context (e.g. personal setting, agencies, medical profession) in which those issues occur, and it is also unclear whether people with intersex/DSD feel that they encounter non-acceptance or discrimination. There is also a lack of clarity regarding participation in education, work and leisure activities. To what extent is there evidence of reduced participation, higher dropout rates, lower performance and a less enjoyable experience of work, education and leisure activities? Do people with intersex/DSD make different choices as regards work, education and leisure activities because of their condition? Is reduced participation, if it occurs, primarily the result of condition-specific characteristics or of actual or anticipated negative reactions? Are there differences between people who have and have not undergone medical interventions?

Research among LGB persons has shown that contact with other LGB persons can help improve psychosocial health (Kuyper 2011; Meyer 2003); in similar vein, research could look at whether there is a comparable relationship for persons with intersex/DSD. This exploratory study shows that patient organisations meet an important need on the part of interviewees with intersex/DSD, but it was not possible to compare people who do and do not have experience of this. It also remained unclear whether everyone with intersex/DSD is aware of the existence of these organisations, has equal access to them, has a need for contact with others, and to what extent they would value such contact. There is a clear medical side to intersex/DSD, though the nature and seriousness of medical interventions varies widely between conditions. Patient-based studies have therefore focused on the medical history of those concerned and what their experience has been. Yet research from a non-patient-based perspective could have clear added value. The first question is to what extent patient-based research is able to provide a complete picture of the experiences of the medical profession by people with intersex/DSD, because they are dependent on doctors and may give a distorted picture when reporting negative incidents. The group that participates in the research may also be specific (see § A.3.3). Moreover, patient-based research generally asks people to state retrospectively how they experienced interventions.[2] Less attention is given to people's views on medical interventions in relation to present and future practice.[3] Asking people with intersex/DSD about this is a sensitive matter, but this information could lead to important insights.

A.3   Methodological issues

One condition for carrying out large-scale survey research among people with intersex/DSD is that the target group is clearly classified and that the entire target group has a known probability of participating in the survey. If this is not the case, the outcomes may present a picture that is distorted or which represents a selection of the target group. Necessary conditions for large-scale research are therefore classification of the target group, identifying those persons and ascertaining who might potentially be recruited to take part. In carrying out survey research among persons with intersex/DSD, classifying the target group and obtaining a large enough non-selective sample[4] is important. This is discussed further below, along with the challenges and opportunities. We would stress that some of the challenges described here are typical of research among small populations, hidden groups and minorities, and as such are not exceptional.

A.3.1 Delimitation and classification of the target group

The most logical way of identifying the target group is to start with the classification used in the Consensus Statement (Hughes et al. 2006). That classification is now commonly accepted among Dutch medical professionals and interest groups (§ 2.3), though in practice there is a slight grey area, for example regarding when hypospadias should be considered 'serious' enough to be classed under DSD. There are also discussions within some patient organisations as to whether their condition should be categorised as intersex/DSD. They may for example distance themselves from other intersex/DSD conditions (which in their view are essentially different), or there may be sensitivities regarding the use of the terms 'intersex' or 'DSD'. Despite wide differences between conditions, however, there are a number of shared problems, points of contact and interests. In the present phase of orientation and scoping out the problem, it is recommended that the target group be classified broadly and that no conditions should be excluded in advance. For reasons of practicality, it makes sense to focus quantitative survey research in the target group on adults with intersex/DSD. Adolescents aged under 16 would require their parents' permission to take part, and that is a sensitive area, partly because parents sometimes have difficulty in accepting their child's condition. Certain conditions and problems moreover only come to light in (early) adulthood, and a survey focusing (exclusively) on children and adolescents would fail to cover certain specific problems relating to social situation. At the same time, it is important to gain more knowledge about the experiences of children and adolescents with intersex/DSD and of their parents, because such information is currently very scarce (COC 2013). There was little scope to elicit such information within this exploratory study, though we devote more attention to this group in section A.5.1.

A.3.2 Selectivity: hidden populations, exclusion and nonresponse

In terms of survey research, the intersex/DSD population can be described as a hidden population. 'Hidden' here means that it is impossible or at least very inefficient to recruit a good sample of respondents from the population with intersex/DSD. There are no complete administrative records in which all persons with intersex/DSD in the Netherlands are registered as such. A national survey in which screening questions are included to determine whether someone has intersex/DSD would be possible, but would require a large sample given the estimated prevalence of intersex/DSD of 0.5% of the total population (§ 2.5).

Exclusion relates to people who belong to the population with intersex/DSD but who have not been identified as such and who can therefore not be invited to take part in research. Examples include people with intersex/DSD who cannot be traced via the recruitment channels used. If people are excluded, this has an impact on the potential composition of the respondent group, and that can distort the outcomes of the research if these persons have different experiences and opinions in relation to key themes. By no means everyone who is known to have intersex/DSD and who is eligible for invitation to participate will be willing to take part in a survey. These persons are referred to as nonrespondents. If these nonrespondents do not share the opinions of those who do take part in the survey, that is a problem. It is possible that people will be less willing to participate if they have never had any unfortunate experiences (and participation in the study is therefore of little importance for them personally), if by contrast they have had very negative experiences (participation is too painful) or if they have great difficulty in coming to terms with their own condition (participation is too confrontational). Nonresponse may also be expected from people who do not readily associate themselves with intersex/DSD or with their own condition.

The likelihood of nonresponse is sometimes greater among certain groups. Discussions with professional experts in this exploratory study revealed that, in addition to children and adolescents (§ 5.5.1), two groups are likely to be less willing to take part in research. The first is men. This is a general phenomenon in survey research, but may play an even bigger role here because, in parallel with homosexuality and gender non-conformity, issues relating to sex and genitalia are extraordinarily sensitive areas for men. The patient organisations for specific intersex/DSD conditions that focus on men appear to have less of a public profile than organisations for conditions which occur (mainly) in women. Whether the embarrassment or taboo is greater for men than for women would need to be determined through research.

Another subcategory where the willingness to participate is likely to be lower is migrants. This again is a general phenomenon in survey research (see e.g. Kappelhof 2010), but there is an additional factor here in that themes relating to sex, gender, sexuality and infertility are sensitive subjects in certain migrant groups. If these groups do not participate in the survey, this could affect the outcomes because it is known from international studies and discussions with Dutch medical professionals that some types of intersex/DSD occur more commonly in migrant groups due to genetic variations and the larger number of intrafamilial marriages.

Finally, it is important that the composition of the respondent group corresponds with the composition of the population with intersex/DSD conditions. If a condition occurs relatively commonly in the Netherlands, but hardly any persons with this condition take part in the survey, the outcomes are likely to produce a distorted picture for the total population with intersex/DSD, especially if the nature and intensity of the problems experienced varies widely between the different conditions. As this is a hidden population, it is not possible to correct for these outcomes using weighting. The composition of the population is unknown, and this makes it all the more important to try to recruit respondents in such a way as to minimise possible selectivity of the respondent group.

A.3.3 Advantages and disadvantages of different recruitment methods for survey research

The foregoing makes clear that the composition of the respondent group is an important factor in obtaining reliable results. Earlier Dutch research among patients with intersex/DSD showed that recruiting respondents can be difficult, partly because of the sensitive themes involved (Callens et al. 2012a). Despite this, Sanches and Wiegers succeeded after considerable effort in obtaining a high response rate in their survey of children and adolescents with congenital adrenal hyperplasia (CAH) (2010). Below we describe the experiences gained with four possible recruitment methods in hidden populations, as well as who might be reached and what the limitations are for the anticipated respondent group. We look in turn at patient-based sampling, sampling via self-referral (convenience sample), respondent-driven sampling (RDS) and sampling via a panel.

Patient-based sampling

Patient-based sampling is the usual method used in research by medical centres specialising in DSD. It entails recruiting survey participants from among the centres' own patients, possibly enlarged to include patients from other specialist medical centres (but rarely regional hospitals or general practitioners). Sometimes respondents are recruited via (a limited number of) patient organisations. In patient-based sampling, potential participants are typically contacted and approached on the basis of their status as patients. This has the advantage of making it possible to determine medically whether participants have a form of DSD, so that only diagnosed patients are included in the research sample. The composition of the patient population is known, and it is therefore also known whether the ultimate composition of the respondent group is a good reflection of that population.

A first disadvantage of patient-based sampling is that not everyone with intersex/DSD is known to medical centres or can be found. Although there are administrative records of patients with intersex/DSD, they will clearly not include everyone with intersex/DSD. Specialist medical centres have several hundred patients with intersex/DSD in their records, whereas according to the prevalence indication (§ 2.5) it is estimated that there are around 80,000 persons in the Netherlands with intersex/DSD. In the first place, this method rules out people who are not aware of their condition. It is known, for example, that a considerable proportion of boys and men with Klinefelter syndrome are not aware that they have the condition. This does not mean they do not have any complaints, but rather that they have not received a correct diagnosis. It is also known that some conditions are difficult to diagnose precisely, which means that people are not always recognised as having intersex/DSD. Another group that is ruled out are people who no longer require medical treatment and who are completely off the radar of medical centres and patient organisations. Older persons may also not be familiar with the current names given to conditions, and people may not feel drawn to respond to calls to take part in research.

A second drawback relates to the probability of selective nonresponse. This sampling method involves approaching persons with intersex/DSD from a medical perspective, leading to the possibility that people who take different views on the research topics will not wish to participate. Examples might include people who have had traumatic experiences of medical interventions (e.g. lots of doctors around their bed, photographs being taken of their genitals, being regarded as an 'interesting case'). A second group comprises persons who do not see themselves as patients and who do not wish to be approached as such. Research by Sanches and Wiegers (2010) revealed that using terms such as 'patients' and 'illness' is a delicate matter, and in their case led to nonresponse. People who have developed a fear of hospitals or who are unwilling to undergo medical diagnosis or tests (e.g. blood screening, gynaecological examination, chromosomal testing, ultrasound or bone scan) will be more difficult to recruit using a patient-based sampling method. This may lead to distortion because the participants and non-participants differ as regards the research questions.

Summarising, the greatest challenges in using a patient-based sampling method for survey research among people with intersex/DSD lie in selective nonresponse and the exclusion of specific intersex/DSD subgroups.

Convenience sampling

Convenience sampling is a method in which people generally take part in the survey because they have registered an interest themselves. In order to obtain a large and diverse respondent group, this sampling method uses a wide range of different recruitment channels, such as newspapers, magazines, television and social media. The snowball method can also be used, asking people to ask others who are members of the target group to take part in the survey. The advantage of this method is that, compared with exclusively patient-based sampling, being a patient is not the most important criterion. This can reduce the number of medically related refusals, thereby improving the selectivity of the respondent group.

On the other hand, it is possible that some people will not be reached by this method, either. For example, it is logical when using convenience sampling among people with intersex/DSD to contact potential participants initially via patient organisations, but not everyone is affiliated to such an organisation. Some do not consider themselves patients, do not or no longer experience major problems, or there is quite simply no patient organisation for their specific condition. The possibility can moreover not be ruled out that people will take part in the survey whose precise intersex/DSD condition has not been medically established. There are for example transgender persons who regard themselves as having an intersex condition, whereas according to the medical definition they do not form part of this group. In this case, therefore, the respondent group consists of respondents who see themselves as someone with an intersex/DSD condition.

It is also advisable to include medical centres in recruitment efforts involving convenience sampling by using posters, flyers and having doctors alert patients. The privacy of patients needs to be taken into account here. It is unclear to what extent the consent of each individual medical centre and the approval of a medical ethics committee for each medical centre is necessary for passive or active recruitment. There is no consensus of views on this, and European directives offer little help. If consent and approval is required for each medical centre, this will slow down and impede recruitment via these channels. In a study carried out by the Netherlands Institute for Health Services Research (Nivel) among young patients with CAH, several medical centres were found to be willing to inform parents of the study by letter during consultations (Sanches & Wiegers 2010). Finally, it should be noted that there is no nonresponse in convenient sampling because all respondents volunteer to participate. It is however possible that certain persons will register earlier than others, possibly for reasons related to the subject of the research. Summarising, the biggest challenges in using this recruitment method for people with intersex/DSD lie in reaching certain groups, the relationship between those who register and the research topic, and the possible inclusion of people who do not form part of the intersex/DSD population according to the generally accepted classification.

Respondent-driven sampling (RDS)

A third recruitment method is respondent-driven sampling (RDS) (Heckathorn 1997; Salganik & Heckathorn 2004; Volz & Heckathorn 2008). RDS is an advanced snowball method which can sometimes be used in survey research involving hidden populations. RDS was for example used in a recent survey of transgender persons (Keuzenkamp 2012). In RDS, recruitment takes place through the networks of participants in the study and takes account of information about those networks.[5] If the conditions are met,[6] the big advantage of this recruitment method is that it is possible to correct for the influence of respondent group selectivity on the outcomes of the survey. In the case of intersex/DSD, however, there are lots of different patient organisations for specific conditions, but outside this circuit people do not appear to be in regular contact with each other. There are shared problems, but no strong identification with others. This makes it unlikely that the conditions for RDS will be met, making this a less obvious recruitment method for this target group.

Recruitment via a panel

The final method discussed here is recruitment via a panel. With this method, a large, existing panel is asked a screening question ('are you or do you know someone...') in order to identify people who belong to the target group. This group are then asked if they are willing to participate in a study on a specific theme. In principle, recruitment via a panel is no different from convenience sampling, though has the potential advantage that there is little connection between the reason for participating in the panel and having intersex/DSD. Thus, the selectivity of the potential respondent group may be lower than with the previous recruitment methods. If people do belong to the target group but do not wish to take part, it is sometimes possible to correct for this to some extent, because background information is often available on the panel members. Self-selection and selective nonresponse remain a challenge with this method. Recruitment via a large panel can deliver a large number of participants, but probably insufficient to rely on this method alone.

The solution? A combination

In order to reach the largest possible number of potential participants within the entire population with intersex/DSD, a combination of recruitment methods would seem to be the most appropriate approach. This was also the approach used in the study of transgender persons by Keuzenkamp (2012). For people with intersex/DSD, consideration could be given to a combination of convenience sampling and recruitment via patient organisations, medical channels, general channels, the snowball method and possibly recruitment via a large, existing panel.

A.4   Practical options for large-scale quantitative research

At the start of this exploratory study, the initial thinking concerning the possibility of investigating the social situation of people with intersex/DSD through large-scale quantitative research was to set up a new survey. This is the first option discussed below.

As the exploratory study progressed, however, two options emerged for tapping into ongoing and existing patient-based research. These options are also described below. Finally, we make an assessment as to which is the best of the three options.

A.4.1 Setting up new research

An advantage of setting up new research is that respondents can be asked very specifically and in great detail about themes relating to their social situation, such as those referred to in section A.2. Moreover, an approach that is not focused on patients could persuade people who do not consider themselves patients, are not willing to undergo medical tests or who have developed a fear of hospitals to take part, creating a less selective sample population.

When recruiting participants for new research, a combination of recruitment methods will lead to the optimum respondent group with a view to minimising selectivity. Based on the exploratory study, for which contact was made with a number of patient organisations, we anticipate that they would be willing to help in the dissemination process. Related channels could also be addressed, focusing on issues relating to infertility and transgender persons, because these networks also contain people with intersex/DSD. The willingness of medical centres to cooperate was not explored as part of this study. If each centre has to give its consent and the approval of medical ethics committee is required for every medical centre, recruitment via medical centres will be very time-consuming and complex.

If a new study is set up using multiple recruitment channels, the size of the respondent group is likely to be higher than with patient-based studies. There is also a strong chance that less specific groups will be ruled out. Earlier research among young LGB persons recruited via convenience samples, for example, showed that general channels such as social media and websites do not generate the majority of participants, but that they do reach subgroups more effectively, for example young bisexuals and LGB persons who have yet to come out (Van Lisdonk & Van Bergen 2010). A large panel drawn from the Dutch population could also provide a useful additional pool of potential respondents. The composition of the respondent group could thus be of higher quality than in a patient-based survey, but there will still be some selectivity, for example due to self-referral or because not every member of the target population knows they have the condition. The respondent group will consist of people who regard themselves as having an intersex condition or DSD (or a specific form of it), rather than people who have been medically diagnosed with the condition.

Finally, it must be borne in mind that the target group is of limited size and that a small group must not be subjected to too much questioning by researchers (Callens 2014). On the other hand, we expect a proportion of the potential participants to show a great willingness to participate, because there has to date been little opportunity for them to make known their experiences in relation to their social situation in research. The impression we gained during the exploratory study was that, while it is not easy to track down people with intersex/DSD, those who took part did so very willingly and fully endorsed the importance of research into and more knowledge of intersex/DSD.

A.4.2 Data analysis on current European study dsd-LIFE

Preparations began in the second half of 2013 for the implementation of the European study dsd-LIFE, which involves research in six countries (including the Netherlands) from a medical, psychological and ethical perspective among people with intersex/DSD. All types of DSD are included in the research. The objectives are as follows:[7]
  - To improve the treatment and care of people with intersex/DSD. The study devotes a good deal of attention to the long-term effects on health, quality of life, sexual functioning and psychological well-being, with the principal purpose of developing European guidelines for treatment and care.
  - To explore how satisfied patients are with their treatment, as well as to solicit their views on how society (and especially medical professionals) respond to their disorder.
  - To provide information and education on the needs and care of these persons.

The study is being carried out in the Netherlands by Radboud University Medical Centre (Radboudumc) and VU University Medical Centre Amsterdam (VUMC). Erasmus MC has been informed of the study and has pledged its cooperation as long as this does not interfere with its own research. Respondent recruitment is taking place in 2014. The survey comprises written questionnaires, an interview about the medical aspects of the condition, and medical tests. Participants must be at least 16 years old. Recruitment will take place exclusively via a number of specialist medical centres, supplemented by communications via patient associations. The study is thus aimed primarily at patients who are known to the specialist medical centres. The envisaged number of participants for the entire European study is estimated at around 1,000.

The questionnaires cover a number of aspects in relation to respondents' social situation. There are items on openness to others, stigma and embarrassment, the social network, perceived acceptance and support, psychosocial well-being, treatment by doctors and ethical issues. One advantage of this study is that as far as possible it uses internationally validated scales, which means that the questions on these topics have been asked several times before, thus allowing comparison with other studies. These scales are often general in nature and are not aimed specifically at the experiences of people with intersex/DSD.

The data will become available for analysis in early 2015. It seems unlikely that the participating Dutch medical centres will publish detailed reports on the social themes covered in the questionnaire. There would appear to be scope for collaboration between medical centres and social science research institutes in reporting on these aspects.

Dsd-LIFE offers the opportunity to present a brief outline of the social situation of patients with intersex/DSD in the Netherlands, but a number of themes related to their social situation that have been identified in this exploratory study will not be covered in any detail. The background to and consequences of people not being open about their condition, adapting their behaviour, impediments to participation and well-being related to stigmatisation, secrecy and (possible) negative reactions from others will for example be left out of consideration.

Another disadvantage is the risk of a selective respondent group, because the respondent group will be restricted mainly to patients. Some groups will therefore be ruled out,
which means the outcomes could present a distorted picture of the population. Moreover, the fact that the research is being carried out by medical centres, with a strong focus
on medical/psychological aspects, could lead to selective nonresponse, which again
could result in a selective respondent group and potentially distorted outcomes (see §
A.3.3).

Although this study has a strong medical/psychological focus, the possibility is being
kept open that a non-medical research institute could contribute to the recruitment of
participants. This could be an interesting opportunity, because it would generate attention for this research via other, non-medical channels as well. A social science research
institute can position itself differently and can persuade participants who are willing to
complete a questionnaire but who prefer to stay away from the medical world and are
unwilling to undergo extensive medical examination. There is also a potential pool of
participants to be recruited who are aware of their condition but who are not affiliated
to medical centres or the limited number of patient organisations. They could complete
only the written questionnaires, for example. This would enlarge the respondent group
and possibly make it less selective.

A.4.3 Secondary data analysis on first large Dutch DSD study

The first large-scale Dutch DSD study involving a large number of patients with DSD was
carried out between 2007 and 2010. It concentrated on the long-term outcomes of sex
development, psychosocial well-being and the physical and psychological development
of women and men with DSD. The focus was on both medical and psychological aspects
in relation to treatments, external appearance, functioning and sexuality. The study was
carried out in collaboration between Erasmus MC, Radboudumc and VUMC, and was
coordinated by Erasmus MC.

Recruitment took place via three medical centres. A total of 109 women and 14 men with
DSD took part. Patients aged under 14 years, patients with non-mosaicism forms of Turner or Klinefelter syndrome and patients with intellectual limitations were not eligible to
participate in the study. This also applied for people whose medical situation had not yet
been precisely established and who did not give their consent to further medical investigation at the time of the study. Initially, the study also included children and adolescents. However, the researchers failed almost entirely to recruit adolescents, as parents
were not willing to allow their children to participate in such a study.

In addition to medical follow-up, psychological questionnaires were distributed in this
study and semi-structured interviews held with participants. The medical follow-up
was focused on the current physical situation, gynaecological or urological complaints,
cosmetic aspects and sexuality. The psychological questionnaires covered gender
development, sexuality and psychosocial well-being. Themes addressed included the
degree of openness towards various people, quality of life, behavioural and emotional
problems, illness cognition, evaluation of information that patients had received about
their condition, satisfaction with their treatment and ethical issues. The semi-structured

interviews dealt among other things with the patient's own experiences of DSD, the reactions of others, psychosexual functioning, social adaptation, self-image and coping with infertility (Callens et al. 2012a; Callens et al. 2012b). It is not known how extensive the questioning on these themes was. Publications have already appeared about this study (including several articles in the dissertations of Callens (2014) and Van der Zwan (2013)).[8] The drawbacks of this option are comparable with those of the dsd-LIFE study. Not all relevant themes are covered (extensively), such as participation and the impact on social relationships. The patient-based recruitment of respondents leads to some exclusion and creates the risk of selective nonresponse, possibly leading to distorted outcomes. This latter point is acknowledged by the researchers (Callens 2014). They did however establish that nonrespondents did not differ in terms of sociodemographic characteristics and type of condition from people who did participate. To date, little attention has been devoted to themes related to social situation, though the researchers are open to this (see note 1). The possibility of performing secondary data analysis focusing on these themes could be explored. The reporting would in that case in all probability be carried out in collaboration with Erasmus MC; which is where the researchers who conducted the interviews are based. An advantage is that the data will be available in the very near future.

A.4.4 Considerations and conclusion regarding the options for large-scale quantitative research

The best option for large-scale quantitative research would appear to be to set up new survey research, using a combination of recruitment methods and with an emphasis on the non-medical aspects. It is however crucial that medical centres are willing to cooperate in the recruitment process. If all recruitment channels can be exploited to the full, the respondent group will be larger and more diverse than in the two patient-based studies. Moreover, adequate attention will then be devoted to themes related to respondents' social situation. There is also likely to be less distortion of outcomes in relation to experiences with medical practice. A non-selective respondent group is not achievable in any study of this target group. That is not uncommon in survey research. The main thing is that an estimate can be made of what the limitations are, so that the reporting can take account of the scope of the outcomes. In this exploratory study, we have been able to identify this reasonably well.

New research is also the most expensive option, because all phases have to be designed and carried out, including an extensive recruitment phase. This is not the case for the other two variants, in which the activities would be limited to carrying out and reporting on additional analyses. The throughput time of the other two variants is also shorter than for new research.

If less importance is attached to the situation of non-patients and an extensive study of their social situation, it would be sufficient to make a selection of themes; a 'light' version based on analysis of the current dsd-LIFE or the first Dutch DSD study would then be an option. In that case, themes such as participation and the impact on social relationships would receive little attention. In the case of dsd-LIFE, a more serious exploration could also be conducted into how additional participants outside the medical circuit could be recruited; the recruitment of participants for this study is still under way. Finally, we would observe that analyses of survey data will be limited mainly to descriptive analyses because of the expected number of participants. The analyses and reporting will also have to take account of the fact that the experiences of people with different conditions can vary widely.

A.5   Other suggestions for research

Although the Ministry only asked for an inventory of the options for large-scale quantitative research, a number of other suggestions for research emerged during this exploratory study which we would mention here.

A.5.1 Research among children

First, several professional experts pointed out the desirability of research among children with intersex/DSD and their parents. Children and their parents usually have recent experiences with medical practice and are often still looking for a way of coming to terms with their condition. The report by COC on the experiences and rights of LGBTI children referred to the lack of research among children with intersex/DSD, whereas there are indications that the situation of these children may be a cause for concern (2013). In the Netherlands, apart from the study of children and adolescents with CAH (Sanches & Wiegers 2010), we are not aware of any other national research among children with intersex/DSD. Undoubtedly, the fact that carrying out research among young people is difficult will play a role here. As stated earlier, researchers who carried out the first Dutch study on DSD commented that young people are often unwilling to take part and that their parents are also reluctant. The voices of children and parents are also absent from this exploratory study, though the interviews with professional experts made clear the importance of gaining more knowledge about their experiences.
If research among children and adolescents is considered, the relationship with parents would need to be a central theme. Several sources suggest that the interests of parents and children can be at odds. With some conditions, decisions are taken whilst the child is very young on medically irreversible interventions.
In order to protect the physical integrity of the child, the Council of Europe recently requested member states to carry out research on the situation of children with intersex (2013). Attention should be given to unnecessary medical or surgical interventions, self-determination for these children and the provision of adequate care to the children and their families (Council of Europe 2013).
Given that even less is known about the experiences of children and adolescents with intersex DSD than for adults, and given that the potential sensitivities can be greater, consideration could be given to an exploratory qualitative study among children and adolescents, their parents and professional experts by means of individual discussions and focus group sessions. Patient organisations and medical centres would be the most logical recruitment channels for coming into contact with children, adolescents and their parents.

A.5.2 A closer look at medical ethics and legal issues

A number of issues relating to medical ethics are mentioned in chapters 3 and 4 (including genital surgery and prenatal diagnosis), as well as in Appendix C. Although these themes were briefly covered in the exploratory study, however, there was no scope for a thorough investigation of them. More in-depth qualitative research could for example look at the specific experiences and views of people who encounter them. Whilst this study was being carried out, Utrecht University began a study on the legal possibilities and consequences of registering a person's sex as indeterminate. The report on this study will be published later in 2014.

A.6   Conclusion

This appendix describes the conditions for large-scale quantitative research into the social situation of persons with intersex/DSD, as well as a number of ways in which such research could be carried out. It gives suggestions for classifying the target group and the content of the research. It makes clear that attention needs to be given to the recruitment of potential participants and that selectivity of the respondent group cannot be ruled out. Whilst taking into account the limitation in relation to the selectivity of the respondent group, large-scale quantitative research is regarded as feasible. Three practical options for carrying out the research are described, as well as the advantages and disadvantages of each.

Appendix B Prevalence table for intersex/DSD[a]

	source and explanation of calculation	conservative prevalence[b]	broader prevalence[b]
Intersex/DSD with ambiguous genitals			
including CAH, PAIS, partial gonadal dysgenesis and partial forms of testosterone biosynthesis defects (e.g. 5alpha-RD2 and 17beta-HSD3)	Source: Hughes et al. 2006. This is 1:4500. This prevalence is generally accepted by Dutch and international scientists, the NNID and a number of patient organisations.	0.022	0.022
other intersex/DSD conditions			
CAIS	Source: Boehmer et al. 1999. Range AIS between 1:40,800-1:99,000, incidence in the Netherlands. 9/10 of these prevalences have been used, since around 10% relate to PAIS (Blackless et al. 2000) and this condition is included in the calculation of DSD with ambiguous genitals.	0.0009	0.002206
XY-gonadal dysgenesis (Swyer syndrome)	Source: Michala et al. 2008, 1:80,000 people. Used by the NNID. NB: www.erfelijkheid.nl assumes 1:30,000 people based on http://ghr.nlm.nih.gov/condition/swyer-syndrome. As the original source is not given and could therefore not be consulted, this prevalence is left out of consideration.	0.001250	0.001250
MRKH	Source of conservative prevalence: Aittomaki et al. 2001, 1:5,000 girls, incidence in Finland. This prevalence is also cited by Dutch medical professionals. Source of broader prevalence: Folch et al. 2000, range between 1:4,000-1:5,000 girls.	0.0100	0.011111
hypospadias	Source: Pierik et al. 2002, 26/10,000 for newborn babies (boys + girls), incidence of hypospadias in Rotterdam with the exception of hypospadias in the region of the glans, as too mild. The total including mild forms is 38/10,000.	0.2600	0.2600

	source and explanation of calculation	conser-vative prevalence[b]	broader prevalence[b]
micropenis	Left out of consideration due to too much over-lap with other conditions.	X	X
47, XXY (Klinefelter syndrome)	Source: Gooren en De Ronde 2006, between 1:600-1:700 boys in the Netherlands, half of whom are diagnosed. Source of broader prevalence: Morris et al. 2008, 1:580 boys. The patient organisation and a number of Dutch medical professionals cite a prevalence of 1:500.	0.0714	0.0862
45, X (Turner syndrome)	Source of conservative prevalence: 1:2,500 girls is a well- known prevalence e.g. Callens 2014; international studies, http://turners. nichd.nih.gov/; http://www.erasmusmc.nl/ alkg-cs/3469011/1297367/4041764/UitlegTurner; lower limit at http://www.erasmusmc.nl/huge/5 1023/177434/2224931/2892534/3433622. Source of broader prevalence: Gravholt 2005, 50:100,000 liveborn girls, incidence in Denmark.	0.020	0.0250
47, XYY	Source: Morris et al. 2008, 1:1,000 boys	0.050	0.050
47, XXX	Source: Morris et al. 2008, 1:1,000 girls	0.050	0.050
total prevalence (%)		0.4858	0.5078
number of persons in the Dutch population (%*16,800,000)		81,616	85,306

Source: SCP
a Assumptions: only liveborn infants; prevalence is assumed to remain unchanged over time (to allow for publication date and generational effects); prevalences in Dutch studies take precedence over international studies; prevalences in international publications are assumed to be comparable to the Netherlands; for calculating the prevalence in the population, it is assumed that the sex ratio of men and women is 50/50; it is assumed that persons do not have two types of intersex/DSD, and prevalences are added together; in the event of doubt about the reliability of sources, prevalences are not included.
b Shown as a percentage of the Dutch population.

Appendix C Attention of organisations for human rights issues in relation to intersex/DSD

In recent years, the attention drawn by human rights organisations and advocates to intersex/DSD in relation to human rights violations has been stepped up. Organisations such as the Council of Europe (2013), the UN Special Rapporteur (Méndez, United Nations 2013), foreign ministers in the Council of the European Union (2013) and the Commissioner for Human Rights of the Council of Europe (2014) are taking the calls by advocates and activists such as the European branch of the International Lesbian, Gay, Bisexual, Trans and Intersex Association (ILGA-Europe) and Organisation Intersex International Europe (OII Europe) increasingly seriously and have recently focused attention on the human rights position of people with intersex/DSD. We give three examples here.

In the non-binding declaration of intent that was signed on the International Day Against Homophobia and Transphobia in May 2014 by representatives of 17 European countries, including the Dutch Minister of Education, Culture and Science, the signatories declared the following: 'We welcome international initiatives taken to promote human rights of LGBTI persons.'

The signatories expressed the following intentions, among others:

> 8 welcome international initiatives aimed at increasing the level of knowledge on the human rights situation of intersex persons;
> 10 adopt measures to promote equality of LGBTI persons at the national level, and support the adoption of strategies at the regional and international level when appropriate;
> 12 cooperate with and consult non-governmental organisations defending the human rights of lesbian, gay, bisexual, trans and intersex persons on the adoption and implementation of measures that may have an impact on the human rights of LGBTI persons and ensure that all civil society actors working on LGBTI rights issues are afforded a safe and enabling environment, in law and practice, to carry out their legitimate functions and operate free from hindrance and insecurity
> (Declaration of Intent 2014)

The Council of Europe has passed a resolution requesting that member states:

> Undertake further research to increase knowledge about the specific situation of intersex people, ensure that no-one is subjected to unnecessary medical or surgical treatment that is cosmetic rather than vital for health during infancy or childhood, guarantee bodily integrity, autonomy and self-determination to persons concerned, and provide families with intersex children with adequate counselling and support. (Council of Europe 2013: 2)

The UN Special Rapporteur on torture and other cruel, inhuman or degrading treatment or punishment has called on all states:

> To repeal any law allowing intrusive and irreversible treatments, including forced genital-normalizing surgery, involuntary sterilization, unethical experimentation, medical display, 'reparative therapies' or 'conversion therapies', when enforced or administered without the free and informed consent of the person concerned. He also calls upon them to outlaw forced

or coerced sterilization in all circumstances and provide special protection to individuals belonging to marginalized groups. (Méndez, United Nations 2013: 23).

Human rights issues in relation to intersex/DSD are focused mainly on two objectives (e.g. Karkazis 2008; Vilain 2006). First, the spotlight is turned on medical practices that affect self-determination, physical integrity and autonomy. Second, there is an attempt to promote the acceptance of intersex/DSD, to embed it in legal frameworks and to change the awareness in society and medical circles. The position of children occupies a special place here; they are a potentially vulnerable group because medical interventions are sometimes carried out at a young age and their autonomy and resilience may be jeopardised (COC 2013; Council of Europe 2013).

Medical ethics aspects in relation to intersex/DSD to which organisations and researchers draw attention include irreversible operations on the external genitals where there is no medical need for the surgery and where the consent of the person concerned is not sought; operations on the internal sex organs resulting in sterilisation; prenatal diagnosis with the potential option of termination of pregnancy; pre-implantation genetic diagnosis (PGD) with potential consequences for embryo selection; prescription of hormone therapy to pregnant women when there is a chance that the baby she is carrying is a girl with CAH; medical interventions and legal proceedings to enhance fertilisation or fulfil the desire for children (e.g. COC 2013; NNID 2013; Karkazis 2008; Commissioner for Human Rights of the Council of Europe 2014; Tamar-Mattis 2013; Wiesemann et al. 2010).

With a view to combating discrimination on grounds of intersex/DSD, advocates draw particular attention to two issues. First, there is a lack of clarity regarding how often and in what areas there is perceived discrimination on the grounds of intersex/DSD, fear of discrimination and negative attitudes in the population. Studies and reports focusing on discrimination mainly conclude that little or no information is available (Agius & Tobler 2012; COC 2013).

The second issue is to what extent people with intersex/DSD are protected by antidiscrimination legislation. The European legal framework seeks to offer protection against discrimination on the grounds of sex, gender identity and gender expression, but it remains unclear how the ground 'sex' should be interpreted and whether people with intersex/DSD fall within it (Agius & Tobler 2012; COC 2013). Agius and Tobler found in 2012 that no legal cases had at that time been brought forward at European level in relation to discrimination on the grounds of intersex/DSD, so that this had not yet been tested (2012). A few proposals and specific examples can be found in other countries of how intersex/DSD can be given a clear place within legal frameworks as a ground for discrimination. Intersex/DSD could for example be included under existing grounds such as sex, gender, sexual orientation and gender identity. A new ground could also be created, as has happened with intersex status in Australia.

Finally, intersex/DSD is linked to the issue of wider opinions in relation to sex registration. The recent change in the law in Germany making it legally possible for a person's sex to be left indeterminate has led to media attention in the Netherlands for the legal framework applied there. The Netherlands Civil Code offers the possibility in cases where there is doubt about the sex of a child to state on the birth certificate that this could not be determined (Book 1 of the Netherlands Civil Code, Section 19d). In practice, doctors will at some point always assign a sex, because it is generally assumed that not doing so will lead to unnecessary stigmatisation. If it is difficult to determine a person's sex and the assigned sex later proves to be incorrect, the Civil Code offers the possibility of changing the registered sex (Book 1, Section 24). The sex stated on the birth certificate can then be changed. A research team from Utrecht University is currently carrying out a study on the legal possibilities and consequences of registering a person's sex as indeterminate. The researchers are investigating what legal and possibly social consequences might ensue from not registering a sex or leaving it indeterminate. Although the issue of registering a person's sex is mainly relevant for transgender persons, legal changes could also have consequences for a (small) proportion of people with intersex/DSD with ambiguous external characteristics or external characteristics that are incongruent with their legally registered sex. There are examples from other countries where the possibility of entering options such as X (sex not stated or indeterminate), FM or MF when registering the sex are being considered or are recognised.

Dutch advocates argue that the Netherlands does not take a leading role when it comes to attention for human rights issues and social visibility in relation to intersex/DSD (COC 2013; NNID 2013). In countries such as Australia, the UK and Germany there is more public debate, visibility and attention from a legal perspective. A number of countries have also set up a commission, council or working group with the aim of formulating standpoints or recommendations on medical ethics issues in relation to intersex/DSD (Deutscher Ethikrat 2012; Swiss National Advisory Commission 2012; Wiesemann et al. 2010). On the other hand, the Netherlands does have patient organisations for a relatively large number of intersex/DSD conditions compared with many other countries, among other things offering those concerned opportunities to increase their resilience or to engage in contact with others with the same condition.

Notes

1   Researchers from Erasmus MC are currently publishing an article on the psychosocial well-being of patients with DSD. The article deals with psychological complaints, self-esteem, quality of life, illness cognition (what people think about their disorder) and fatigue.
2   Research moreover shows that people assess their experiences in rather different ways depending on whether they are questioned about them contemporaneously or retrospectively (Kahneman 2011).
3   The ongoing dsd-LIFE study devotes some attention to this. See also section A.4.2.
4   Respondent group corresponds with the net sample. The gross sample is equal to the number of persons asked to participate. Those who do not participate constitute the nonresponse.
5   Information is gathered on the recruitment chain, for example who is invited to take part and how many persons with the same characteristics (here, for example, the number of persons with the same intersex/DSD condition) know the participant. This information is then used to weight the data collected and to enable a general statement to be made about the entire target group.

6   The conditions which must be met include things such as the ability for everyone in the population to be put in contact with each other via a single network. Members of the population must also be able to recognise each other as belonging to that population.

7   The objectives are taken from the flyer for potential participants, but are reproduced here in more concise form.

8   The data collected on social well-being are currently being worked up and will be published in the near future.

References

Agius, Sylvan and Christa Tobler (2012). *Trans and Intersex people: Discrimination on the grounds of sex, gender identity and gender expression.* Luxembourg: Office for Official Publications of the European Union.

Aittomaki, K., H. Eroila and P. Kajanoja (2001). A population-based study of the incidence of Mullerian aplasia in Finland. In: *Fertility and Sterility*, vol. 76, no. 3, pp. 624-625.

Alderson, J., A. Madill and A. Balen (2004). Fear of devaluation: understanding the experience of intersexed women with androgen insensitivity syndrome. In: *British Journal of Health Psychology*, vol. 9, no. 1, pp. 81-100.

Blackless, Melanie, Anthony Charuvastra, Amanda Derryck, Anne Fausto-Sterling, Karl Lauzanne and Ellen Lee (2000). How Sexually Dimorphic Are we? Review and Synthesis. In: *American Journal of Human Biology*, vol. 12, no. 2, pp. 151-166.

Boehmer, Annemie L.M., Albert O. Brinkman, Lodewijk A. Sandkuijl, Dicky J.J. Halley, Martinus F. Niermeijer, Stefan Andersson, Frank H. de Jong, Hülya Kayserili, Monique A. de Vroede, Barto J. Otten, Catrienus W. Rouwe, Berenice B. Mendoncca, Cidade Rodrigues, Hans H. Bode, Petra E. de Ruiter, Henrietta A. Delemarre-van de Waal and Stenvert L.S. Drop (1999). 17b-Hydroxysteroid Dehydrogenase-3 Deficiency: Diagnosis, Phenotypic Variability, Population Genetics, and Worldwide Distribution of Ancient and de Novo Mutations. In: *The Journal of Clinical Endocrinology & Metabolism*, vol. 82, no. 12, pp. 4713-4721.

BuZa (2013) *Beleidskader Matra CoPROL, tweede fase.* Besluit van de minister van Buitenlandse Zaken van 12 juli 2013 tot vaststelling van beleidsregels en een subsidieplafond voor subsidiëring op grond van de Subsidieregeling ministerie van Buitenlandse Zaken 2006 (Matra CoPROL 2014-2015). Den Haag: Ministerie van Buitenlandse Zaken.

Callens, Nina (2014). *The Past, the present, the future: Genital treatment practices in disorders of sex development under scrutiny* (dissertation). Ghent: Ghent University.

Callens, Nina, Yvonne G. van der Zwan, Stenvert L.S. Drop, Martine Cools, Catharina M. Beerendonk, Katja P. Wolffenbuttel and Arianne B. Dessens (2012a). Do Surgical Interventions Influence Psychosexual and Cosmetic Outcomes in Women with Disorders of Sex Development. In: *International Scholarly Research Network Endocrinology*, vol. 2012, no. 5, pp. 1-8.

Callens, Nina, Griet de Cuypere, Katja P. Woffenbuttel, Catharina C.M. Beerendonk, Yvonne G. van der Zwan, Marjan van den Berg, Stan Monstrey, Maaike E. van Kuyk, Petra De Sutter, Belgian-Dutch Study Group on DSD, Arianne B. Dessens and Martine Cools (2012b). Long-Term Psychological and Anatomical Outcome after Vaginal Dilation or Vaginoplasty: A Comparative Study. In: *The Journal of Sexual Medicine*, vol. 9, no. 7, pp. 1842-1851.

Claahsen-Van der Grinten, E.M. van Kuyk, A.S. Dessens, S.L.S. Drop en B.J. Otten (2008). De pasgeborene met een gestoorde geslachtelijke ontwikkeling. In: *Tijdschrift voor Kindergeneeskunde*, vol. 76, no. 3, pp. 105-112.

COC (2013). *LHBTI-kinderen in Nederland: Rapportage over de leefwereld en rechten van een vergeten groep kwetsbare kinderen.* Amsterdam: COC.

Cohen-Kettenis, P.T. (2010). Psychosocial and psychosexual aspects of disorders of sex development. In: *Best Practice & Research Clinical Endocrinology & Metabolism*, vol. 24, no. 2, pp. 325-334.

Commissioner for Human Rights of the Council of Europe (2014). *A boy or a girl – intersex people lack recognition in Europe* (Human Rights Comment, 9 May 2014). Consulted on 3 June 2014 at www.humanrightscomment.org.

Council of Europe (2013). *Children's right to physical integrity* (Resolution 1952, 2013). Strasbourg: Council of Europe.

Den Dungen, E.G.H.C. van, L.V. Mijnders, B.J. Otten and M.M.L. Stikkelbroeck (2002). *Een volwassen leven, met AGS*. Den Haag: Jan Evers.

Dessens, A.B. and P.T. Cohen-Kettenis. (2008). Genderrol en genderidentiteit bij geslachtsdifferentie-stoornissen: Voorkomen en psychologische behandeling. In: *Tijdschrift voor Kindergeneeskunde,* vol. 76, no. 3, pp. 137-144.

Deutscher Ethikrat (2012). *Intersexualität: Stellungnahme*. Berlin: Deutscher Ethikrat.

European Parliament (2014). *European Parliament resolution of 4 February 2014 on the EU Roadmap against homophobia and discrimination on grounds of sexual orientation and gender identity* (2013/2183(INI)). Consulted on 3 June 2014 at http://www.europarl.europa.eu/sides/getDoc.do?type=TA&reference=P7-TA-2014-0062&language=EN&ring=A7-2014-0009.

European Union (2013). *Guidelines to Promote and Protect the Enjoyment of All Human Rights By Lesbian, Gay, Bisexual, Transgender and Intersex (LGBTI) Persons* (Luxembourg, 24 June 2013). Consulted on 3 June 2014 at www.consilium.europa.eu/uedocs/cms_Data/docs/pressdata/EN/foraff/137584.pdf

EuroPSI (2014). *What is intersex/dsd*. Consulted on 4 February 2014 on http://www.europsi.org/WhatIsIntersex.

Folch, M., I. Pigem and J.C. Konje (2000). Müllerian Agnesis: Etiology, Diagnosis, and Management. In: *Obstetrical and Gynecological Survey,* vol. 55, no. 10, pp. 664-649.

Gooren, L.J.G. and W. de Ronde (2006). Enkele nieuwe aspecten van het Klinefeltersyndroom. In: *Nederlands Tijdschrift voor Geneeskunde,* jg. 150, nr. 49, pp. 2693-2696.

Gravholt, Claus Højbjerg (2005). Epidemiological, Endocrine and Metabolic Features in Turner Syndrome. In: *Arquivos Brasileros de Endocrinologia & Metabologia,* vol. 49, no. 1, pp. 145-156.

Heckathorn, D.D. (1997). Respondent-driven sampling: a new approach to the study of hidden populations. In: *Social problems,* vol. 44, no. 2, pp. 174-199.

Heesch, M. van (forthcoming). *Ze wisten niet of ik een jongen of een meisje was. Kennis en keuze rondom geslachtsvariaties* (dissertation). Amsterdam: University of Amsterdam.

Heesch, Margriet van (2009). Do I Have XY Chromosomes? In: M. Holmes (ed.), *Critical Intersex* (123-145). Surrey: Ashgate Publishing.

Hughes, I.A., C. Houk, S.F. Ahmed and P.A. Lee, LWPES1/ESPE2 Consensus Group (2006). Consensus statement on management of intersex disorders. In: *Archives of Disease in Childhood,* vol. 91, no. 7, pp. 554-562.

Hughes, Ieuan (2010). How should we classify intersex disorders? In: *Journal of Pediatric Urology,* vol. 6, no. 5, pp. 447-448.

IDAHO (2014). *Declaration of Intent*. International Day Against Homophobia and Transphobia (Valetta, 14 May 2014).

ILGA Europe (2013). *Public Statement by the Third International Forum* (Valetta, 2 December 2013). Consulted on 3 June 2014 www.ilga-europe.org/home/news/latest/intersex_forum_2013

Johannsen, Trine H., Carolina P.L. Ripa, Erik L. Mortensen and Katharina M. Main (2006). Quality of life of 70 women with disorders of sex development. In: *European Journal of Endocrinology,* vol. 155, no. 6, pp. 877-885.

Kahneman, D. (2011). *Thinking, fast and slow*. New York: Farrar, Straus and Giroux.

Kalsbeek, C.J.C. and V.J.D. Platteel (2012). *Kwaliteitscriteria zorg Klinefelter Syndroom vanuit patiëntenperspectief*. Amersfoort: Nederlandse Klinefelter Vereniging.

Kappelhof, J. (2010). *Op maat gemaakt? Een evaluatie van enkele responsverbeterende maatregelen onder Nederlanders van niet-westerse afkomst*. Den Haag: Sociaal en Cultureel Planbureau.

Karkazis, Katrina Alicia (2008). *Fixing sex: intersex, medical authority, and lived experience*. Durham: Duke University Press.

Keuzenkamp, Saskia (ed.) (2010). *Steeds gewoner, nooit gewoon. Acceptatie van homoseksualiteit in Nederland*. Den Haag: Sociaal en Cultureel Planbureau.

Keuzenkamp, Saskia (2012). *Worden wie je bent. Het leven van transgenders in Nederland*. Den Haag: Sociaal en Cultureel Planbureau.

Keuzenkamp, Saskia, David Bos, Jan Willem. Duyvendak and Gert Hekma (ed.) (2006). *Gewoon doen. Acceptatie van homoseksualiteit in Nederland*. Den Haag: Sociaal en Cultureel Planbureau.

Keuzenkamp, Saskia (ed.), Niels Kooiman and Jantine van Lisdonk (2012). *Niet te ver uit de kast: ervaringen van homo- en biseksuelen in Nederland*. Den Haag: Sociaal en Cultureel Planbureau.

Kuyper, L. (2011). *Sexual orientation and health. General and minority stress factors explaining health differences between lesbian, gay, bisexual and heterosexual individuals* (dissertation). Utrecht: Utrecht University.

Kuyper, L. (2012). Transgenders in Nederland: Prevalentie en attitude. In: *Tijdschrift voor Seksuologie*, jg. 36, nr. 2, pp. 129-135.

Kuyper, Lisette. (2013). *Seksuele oriëntatie en werk. Ervaringen van lesbische, homoseksuele, biseksuele en heteroseksuele werknemers*. Den Haag: Sociaal en Cultureel Planbureau.

Liao, L.M. (2003). Learning to assist women born with atypical genitalia: journey through ignorance, taboo and dilemma. In: *Journal of Reproductive and Infant Psychology*, vol. 21, no. 3, pp. 229-238.

Liao, Lih-Mei and Mary Boyle (2004a). Intersex (Special Issue). In: *The Psychologist*, vol. 17, no. 8, pp. 446-462.

Liao, Lih-Mei and Mary Boyle (2004b). Surgical feminising: The right approach? In: *The Psychologist*, vol. 17, no. 8, pp. 459-462.

Liao, Lih-Mei and Katrina Roen (2014). Intersex/DSD post-Chicago: new developments and challenges for psychologists. In: *Psychology & Sexuality*, vol. 5, no. 1, pp. 1-4.

Liao, Lih-Mei and Margaret Simmonds (2014). A values-driven and evidence-based health care psychology for diverse sex development. In: *Psychology and Sexuality*, vol. 5, no. 1, pp. 83-101.

Lisdonk, Jantine van, and Diana van Bergen (2010). SameFeelings: een onderzoek onder homojongeren. In: S. Keuzenkamp (ed.), *Steeds gewoner, nooit gewoon. Acceptatie van homoseksualiteit in Nederland* (pp. 121-131). Den Haag: Sociaal en Cultureel Planbureau.

MacKenzie, D., A. Huntington and J.A. Gilmour (2009). The experience of people with an intersex condition: a journey from silence to voice. In: *Journal of Clinical Nursing*, vol. 18, no. 12, pp. 1775-1783.

Marteau, T.M., I. Nippert, S. Hall, C. Limbert, M. Reid and M. Bobrow, A. Cameron, A. Cornel, M. van Diem ,B. Eiben, S. Garcia-Minaur, J. Goujard, D. Kirwan, K. McIntosh, P. Soothill; C. Verschuuren-Bemelmans, C. de Vigan, S. Walkinshaw, L. Abramsky, F. Louwen, P. Miny, J. Horst, DADA Study Group (2002). Outcomes of pregnancies diagnosed with Klinefelter syndrome: the possible influence of health professionals. In: *Prenatal Diagnosis*, vol. 22, no. 7, pp. 562-566.

Méndez, Juan E., United Nations (2013). *Report of the Special Rapporteur on torture and other cruel, inhuman or degrading treatment or punishment*, Consulted on 3 June 2014 at www.ohchr.org/Documents/HRBodies/ HRCouncil/RegularSession/Session22/A.HRC.22.53_English.pdf

Meyer, I.H. (2003). Prejudice, social stress, and mental health in lesbian, gay, and bisexual populations: Conceptual issues and research evidence. In: *Psychological Bulletin*, vol. 129, no. 5, pp. 674-697.

Meyer-Bahlburg, H.F., C. Dolezal, S.W. Baker and M.I. New (2008). Sexual orientation in women with classical or non-classical congenital adrenal hyperplasia as a function of degree of prenatal androgen excess. In: *Archives of Sexual Behavior*, vol. 37, no. 1, pp. 85-99.

Michala, L., D. Goswani, S. Creighton and G. Conway (2008). Swyer syndrome: presentation and outcomes. *BJOG: An International Journal of Obstetrics & Gynaecology*, vol. 115, no. 6, pp. 737-741.

Morland, Iain (2004). Thinking with the phallus. In: *The Psychologist*, vol. 17, no. 8, pp. 448-450.

Morris, Joan K., Eva Alberman, Claire Scott and Patricia Jacobs (2008). Is the prevalence of Klinefelter syndrome increasing? In: *European Journal of Human Genetics*, vol. 16, pp. 163-170.

NNID (2013). *Standpunten & Beleid 2013/2014*. Consulted on 8 May 2014 at www.nnid.nl/beleid

Pierik, Frank H., Alex Burdorf, J.M. Rien Nijman, Sabine M.P.F. de Muinck Keizer-Schrama, R.E. Juttmann and Robertus F.A. Weber (2002). A high hypospadias rate in The Netherlands. In: *Human Reproduction*, vol. 17, no. 4, pp. 1112-1115.

Reis, Elizabeth (2007). Divergence or Disorder? The politics of naming intersex. In: *Perspectives in Biology and Medicine*, vol. 50, no. 4, pp. 535-543.

Salganik, M.J. and D.D. Heckathorn (2004). Sampling and Estimation in Hidden Populations Using Respondent Driven Sampling. In: *Sociological methodology*, vol. 34, no. 1, pp. 193-240.

Sanches, S. en T. Wiegers (2010). *Het fysiek, sociaal en maatschappelijk functioneren van kinderen met AGS (en hun ouders)*. Nivel: Utrecht.

Slijper, F.M.E., P.G. Frets, A.L.M. Boehmer, S.L.S. Drop and M.F. Niermeijer (2000). Androgen Insensitivity Syndrome (AIS): Emotional Reactions of Parents and Adult Patients to the Clinical Diagnosis of AIS and Its Confirmation by Androgen Receptor Gene Mutation Analysis. In: *Hormone Research*, vol. 53, no. 1, pp. 9-15.

Streuli, J.C., E. Vayena, Y. Cavicchia-Balmer and J. Huber (2013). Shaping parents: impact of contrasting professional counseling on parents' decision making for children with disorders of sex development. In: *The Journal of Sexual Medicine*, vol. 10, no. 8, pp. 1953-1960.

Swiss National Advisory Commission (2012). *On the management of differences of sex development: Ethical issues relating to "intersexuality"*. Bern: Swiss National Advisory Commission on Biomedical Ethics.

Tamar-Mattis, Anne (2013). Medical Treatment of People with Intersex Conditions as Torture and Cruel, Inhuman, or Degrading Treatment or Punishment. In: *Torture in Healthcare Settings: Reflections on the Special Rapporteur on Torture's 2013 Thematic Report* (pp. 91-104). Consulted on 3 June 2014 at www. Washington: Center for human rights & humanitarian law.

TK (2012/2013). *Hoofdlijnen emancipatiebeleid 2013-2016*. Tweede Kamer, vergaderjaar 2012/2013, 32824, nr. 7.

Vilain, Eric (2006). Genetics of Intersexuality. In: *Journal of Gay & Lesbian Psychotherapy*, vol. 10, no. 2, pp. 9-26.

Volz, E. and D.D. Heckathorn (2008). Probability based estimation theory for respondent driven sampling. In: *Journal of Official Statistics*, vol. 24, no. 1, pp. 79.

Wiesemann, Claudia, Susanne Ude-Koeller, Gernot H.G. Sinnecker and Ute Thyen (2010). Ethical principles and recommendations for the medical management of differences of sex development (DSD)/intersex in children and adolescents. In: *European Journal of Pediatrics*, vol. 169, no. 6, pp. 671-679.

Zwan, Yvonne G. van der (2013). *Disorders of Sex Development: Clinical outcomes, (epi)genetic regulation and germ cell cancer* (dissertation). Rotterdam: Erasmus Universiteit Rotterdam.

Websites

http://turners.nichd.nih.gov/
http://www.erasmusmc.nl/alkg-cs/3469011/1297367/4041764/UitlegTurner
http://www.erasmusmc.nl/huge/51023/177434/2224931/2892534/3433622
http://ghr.nlm.nih.gov/condition/swyer-syndrome
www.erfelijkheid.nl

Publications of the Netherlands Institute for Social Research | scp in English

Sport in the Netherlands (2007). Annet Tiessen-Raaphorst, Koen Breedveld.
ISBN 978 90 377 0302 3

Market Place Europe. Fifty years of public opinion and market integration in the European Union. European Outlook 5 (2007). Paul Dekker, Albert van der Horst, Henk Kox, Arjan Lejour, Bas Straathof, Peter Tammes, Charlotte Wennekers. ISBN 978 90 377 0306 1

Explaining Social Exclusion. A theoretical model tested in the Netherlands (2007). Gerda Jehoel-Gijsbers, Cok Vrooman. ISBN 978 90 377 0325 2

Out in the Netherlands. Acceptance of homosexuality in the Netherlands (2007). Saskia Keuzenkamp, David Bos. ISBN 978 90 377 0324 5

Comparing Care. The care of the elderly in ten EU-countries (2007). Evert Pommer, Isolde Woittiez, John Stevens. ISBN 978 90 377 0303 0

Beyond the breadline (2008). Arjan Soede, Cok Vrooman. ISBN 978 90 377 0371 9

Facts and Figures of the Netherlands. Social and Cultural Trends 1995-2006 (2008). Theo Roes (ed.).
ISBN 978 90 377 0211 8

Self-selection bias versus nonresponse bias in the Perceptions of Mobility survey. A comparison using multiple imputation (2008). Daniel Oberski. ISBN 978 90 377 0343 6

The future of the Dutch public library: ten years on (2008). Frank Huysmans, Carlien Hillebrink.
ISBN 978 90 377 0380 1

Europe's Neighbours. European neighbourhood policy and public opinion on the European Union. European Outlook 6 (2008). Paul Dekker, Albert van der Horst, Suzanne Kok, Lonneke van Noije, Charlotte Wennekers. ISBN 978 90 377 0386 3

Values on a grey scale. Elderly Policy Monitor 2008 (2008). Cretien van Campen (ed.).
ISBN 978 90 377 0392 4

The Netherlands Institute for Social Research | scp at a glance. Summaries of 16 scp-research projects in 2008 (2009). ISBN 978 90 377 0413 6

Sport in the Netherlands (2009). Annet Tiessen-Raaphorst, Koen Breedveld.
ISBN 978 90 377 0428 0

Strategic Europe. Markets and power in 2030 and public opinion on the European Union (2009). Paul Dekker, Albert van der Horst, Paul Koutstaal, Henk Kox, Tom van der Meer, Charlotte Wennekers, Teunis Brosens, Bas Verschoor. ISBN 978 90 377 0440 2

Building Inclusion. Housing and Integration of Ethnic Minorities in the Netherlands (2009). Jeanet Kullberg, Isik Kulu-Glasgow. ISBN 978 90 377 0442 6

Making up the Gap, Migrant Education in the Netherlands (2009). Lex Herweijer.
ISBN 978 90 377 0433 4

Rules of Relief. Institutions of social security, and their impact (2009). J.C. Vrooman.
ISBN 978 90 377 0218 7

Integration in ten trends (2010). Jaco Dagevos and Mérove Gijsberts. ISBN 78 90 377 0472 3

Monitoring acceptance of homosexuality in the Netherlands (2010). Saskia Keuzenkamp.
ISBN 978 90 377 484 6

The minimum agreed upon. Consensual budget standards for the Netherlands (2010). Stella Hoff, Arjan Soede, Cok Vrooman, Corinne van Gaalen, Albert Luten, Sanne Lamers.
ISBN 978 90 377 0472 3

The Social State of the Netherlands 2009 (2010). Rob Bijl, Jeroen Boelhouwer, Evert Pommer, Peggy Schyns (eds.). ISBN 978 90 377 0466 2

At home in the Netherlands. Trends in integration of non-Western migrants. Annual report on Integration 2009 (2010). Mérove Gijsberts and Jaco Dagevos. ISBN 978 90 377 0487 7

In the spotlight: informal care in the Netherlands (2010). Debbie Oudijk, Alice de Boer, Isolde Woittiez, Joost Timmermans, Mirjam de Klerk. ISBN 978 90 377 0497 6

Wellbeing in the Netherlands. The SCP life situation index since 1974 (2010). Jeroen Boelhouwer. ISBN 978 90 377 0345 0

Just different, that's all. Acceptance of homosexuality in the Netherlands (2010). Saskia Keuzenkamp et al. (ed.) ISBN 978 90 377 0502 7

Acceptance of homosexuality in the Netherlands 2011. International comparison, trends and current situation (2011). Saskia Keuzenkamp. ISBN 978 90 377 0580 5

Living together apart. Ethnic concentration in the neighbourhood and ethnic minorities' social contacts and language practices (2011). Miranda Vervoort. ISBN 978 377 0552 2

Frail older persons in the Netherlands. Summary (2011). Cretien van Campen (ed.) ISBN 978 90 377 0563 8

Frail older persons in the Netherlands (2011). Cretien van Campen (ed.) ISBN 978 90 377 0553 9

Measuring and monitoring immigrant integration in Europe (2012). Rob Bijl and Arjen Verweij (eds.) ISBN 978 90 377 0569 0

The Social State of the Netherlands 2011. Summary (2012). Rob Bijl, Jeroen Boelhouwer, Mariëlle Cloïn, Evert Pommer (eds.) ISBN 978 90 377 0605 5

Countries compared on public performance. A study of public sector performance in 28 countries (2012). Jedid-Jah Jonker (ed.) ISBN 978 90 377 0584 3

A day with the Dutch. Time use in the Netherlands and fifteen other European countries (2012). Mariëlle Cloïn. ISBN 978 90 377 0606 2

Acceptance of lesbian, gay, bisexual and transgender individuals in the Netherlands 2013 (2013). Saskia Keuzenkamp and Lisette Kuyper. ISBN 978 90 377 0649 9

Towards Tolerance. Exploring changes and explaining differences in attitudes towards homosexuality in Europe (2013). Lisette Kuyper, Jurjen Iedema, Saskia Keuzenkamp. ISBN 978 90 377 0650 5

Using smartphones in survey research: a multifunctional tool (2013). Nathalie Sonck and Henk Fernee. ISBN 978 90 377 0669 7

Perceived discrimination in the Netherlands (2014). Iris Andriessen, Henk Fernee and Karin Wittebrood. ISBN 978 90 377 0699 4

Who cares in Europe. A comparison of long-term care for the over-50s in sixteen European countries (2014). Debbie Verbeek-Oudijk, Isolde Woittiez, Evelien Eggink and Lisa Putman. ISBN 978 90 377 0681 9.

Living with Intersex/DSD. An exploratory study of the social situation of persons with intersex/DSD (2014). Jantine van Lisdonk. ISBN 978 90 377 0717 5